T0135484

Winfried Säckel

Simulation of Spray Polymerisation and Structure Generation in Spray Drying by Single Droplet Models

Logos Verlag Berlin

Bibliographic information published by Die Deutsche Bibliothek

Die Deutsche Bibliothek lists this publication in the Deutsche Nationalbibliografie;
detailed bibliographic data is available in the Internet at http://dnb.ddb.de.

D93

Logos-Ökobuch.
https://www.logos-verlag.de/oekobuch

Logos Verlag Berlin GmbH
Georg-Knorr-Str. 4, Geb. 10, 12681 Berlin

Tel.:   +49 (0)30 / 42 85 10 90
Fax: +49 (0)30 / 42 85 10 92

https://www.logos-verlag.de

Simulation of Spray Polymerisation and Structure Generation in Spray Drying by Single Droplet Models

Von der Fakultät 4 Energie-, Verfahrens- und Biotechnik
der Universität Stuttgart zur Erlangung der Würde
eines Doktors der Ingenieurwissenschaften (Dr.-Ing.)
genehmigte Abhandlung

Vorgelegt von
Winfried Säckel
geboren in Sande (Friesland)

Hauptberichter:	Prof. Dr.-Ing. Ulrich Nieken
Mitberichter:	Prof. Dr.-Ing. habil. Evangelos Tsotsas

Tag der mündlichen Prüfung: 23.11.2021

Institut für Chemische Verfahrenstechnik
2022

Abstract

Spray Drying is a widely applied industrial process, in which granular material is obtained from a liquid feed. The idea of performing polymerisation reactions within a spray in order to integrate several process steps (polymerisation, drying of a solvent, generation of a spherical product) into one apparatus has been discussed for a long time and has frequently been subject to experimental investigations. A mathematical model of droplet polymerisation, which accounts for transport processes inside the drop and predicts polymer properties, has not been published so far. A second matter in spray drying is the evolution of a product's morphology. Established quasi-homogeneous, spherically symmetric single droplet models presuppose a certain morphology and do not allow for structure simulation on a detailed scale. Both questions are addressed by new numerical models, in the first part by an enhancement of established one-dimensional droplet models for reactive drying, in the second part by a novel approach for the simulation of structure generation using a meshfree method.

Single Droplet Modelling of Spray Polymerisation
Free radical polymerisation within a droplet is considered by both lumped and one-dimensionally resolved models of solution drying with additional chemical reactions. Drying leads to an inhomogeneous distribution of the reactants' mixture and affects chemical reactions. Polymer properties are calculated using the method of moments. Its molecular weight depends on reaction conditions and varies through time and space. A sound mathematical desciption of the reaction-diffusion system is derived and accounts for density changing reactions and the varying specific volume of the polymer component. Moreover, a new approach of polymer moments' diffusion is introduced, which - in contrast to the literature approach - preserves spatial inhomogeneities of polymer properties.

Numerical simulations reveal that drying and chemical reactions are not simultaneous, but mostly sequential mechanisms. After solvent vaporisation, polymerisation is performed at a high monomer content, practically as bulk polymerisation. Evaporation of a volatile monomer will decrease the polymer yield tremendously. Further investigations are conducted on various process and model parameters. Several process variants are discussed by means of numerical DoEs. Polymerisation in solvent is scarcely applicable under most reasonable parameter settings. Bulk polymerisation can provide a feasible yield at elevated monomer saturations of the gas phase. Depending on non-ideal thermodynamics, reactive absorption appears as the main driver for increased polymer production. Slight pre-polymerisation before atomisation may provide an additional, yet limited gain in yield. Comparison between the lumped and the one-dimensional model shows that a simple 0D approximation is only feasible in the simple case of a non-volatile monomer.

Simulation of Structure Development within a Droplet by the Meshfree SPH Method

Structure evolution is a complicated process involving interactions of different phases, various physical effects and different time-scales. The meshfree SPH approach represents the continuum by irregularly distributed interpolation points, which move according to a Lagrangian point of view. When such points belong to different physical phases, the method can describe evolving interfaces and structure evolution efficiently. A novel approach for drying of a slurry is introduced. As SPH has not been applied to droplet drying yet, fundamental effects need to be derived. This involves free surface heat and mass transfer according to linear driving forces and an efficient, implicit solution of the energy equation and various approaches for crust formation in the second drying period. Moreover, a new approach for modelling surface tension by pairwise forces is introduced. By adjustment of model parameters, formation of dense, compact structures as well as hollow granules of different crust porosity can be simulated. These parameters can be interpreted in the sense of the binder content. Finally, a new SPH-grid coupled model is presented, which directly computes diffusion in the gas via an underlying mesh. Through this, the receding of water within a porous structure by drying is calculated.

Zusammenfassung

Die Sprühtrocknung ist ein verbreiteter industrieller Prozess, um granuläres Pulver aus einer flüssigen Vorlage zu erzeugen. Die Grundidee der Polymerisation in einem Spray, um mehrere Prozessschritte in einem Apparat zu vereinen (Polymerisation, Abtrennung des Lösungsmittels, Herstellung sphärischer Partikel aus dem Polymer), wird seit langer Zeit diskutiert und immer wieder experimentell untersucht. Ein mathematisches Modell, welches Transportprozesse innerhalb der Tropfen berücksichtigt und Polymereigenschaften vorhersagt, wurde jedoch bisher nicht veröffentlicht. Eine zweite Fragestellung der Sprühtrocknung betrifft die Ausbildung der Morphologie des finalen Produkts. Etablierte quasi-homogene, auf Kugelsymmetrie basierende Einzeltropfenmodelle setzen eine spezifische Struktur des Produkts voraus und ermöglichen nicht die Simulation der Morphogenese auf einer Detailskala. Beide Fragestellungen werden mit Hilfe neuer numerischer Modell adressiert, im ersten Teil über die Erweiterung etablierter eindimensionaler Tropfenmodelle für reaktive Trocknungsprozesse, im zweiten Teil mit Hilfe eines neuartigen Ansatzes zur Simulation der Strukturausbildung basierend auf einer gitterfreien Methode.

Einzeltropfenmodellierung der Sprühpolymerisation

Freie radikalische Polymerisation innerhalb eines Tropfens wird als konzentriertes und als ortsverteiltes Modell der Lösungstrocknung mit chemischen Reaktionen berücksichtigt. Trocknung beeinflusst die Verteilung der Edukte und die Polymerisationsreaktionen. Polymereigenschaften werden mit Hilfe der Momentenmethode berechnet. Das Molekulargewicht des Polymers hängt von den Reaktionsbedingungen ab und variiert örtlich und zeitlich. Eine konsistente Beschreibung des Reaktions-Diffusions-Systems wird hergeleitet und berücksichtigt dichteverändernde Reaktionen und veränderliche spezifische Volumina des

Polymers. Ein neuer Ansatz zur Diffusion statistischer Momente erhält - anders als der Literaturansatz - örtliche Inhomogenitäten von Polymereigenschaften.

Trocknung und chemische Reaktionen erfolgen nicht gleichzeitig. Nach Verdampfung des Lösungsmittel findet die Polymerisation bei hohen Monomergehalten statt, die praktisch einer Massepolymerisation entsprechen. Durch Verdampfung flüchtiger Monomere nimmt die Ausbeute drastisch ab. Prozess- und Modellparameter der Tropfenpolymerisation werden eingehender untersucht und verschiedene Prozessvarianten mit Hilfe numerischer DoEs diskutiert. Das Prozessfenster zur Lösungspolymerisation ist sehr begrenzt. Massepolymerisation ermöglicht bei erhöhtem Monomergehalt im Gas eine hinreichende Ausbeute, die infolge reaktiver Absorption in Abhängigkeit von der nicht-idealen Thermodynamik auftritt. Eine vorgeschaltete Teil-Polymerisation vor Zerstäubung der Vorlage kann in Grenzen die Ausbeute erhöhen. Ein konzentriertes Modell beschreibt nur den einfachen Fall eines nichtflüchtigen Monomers korrekt.

Simulation der Strukturentwicklung über die gitterfreie SPH-Methode
Strukturausbildung ist ein komplexer Vorgang, der die Interaktion verschiedener Phasen, diverse physikalische Effekte und unterschiedliche Zeitskalen betrifft. Der gitterfreie SPH-Ansatz beschreibt das Kontinuum durch unregelmäßig verteilte Interpolationspunkte, die sich gemäß einer Lagrange-Betrachungsweise bewegen. Werden derartige Punkte verschiedenen physikalische Phasen zugeordnet, lassen sich verändernde Grenzflächen und strukturbildende Prozesse effizient beschreiben. Da SPH bisher nicht in Bezug auf Tropfentrocknung eingesetzt wurde, werden grundlegende Ansätze hergeleitet. Dies betrifft Wärme- und Stofftransport über die Tropfenoberfläche mit Hilfe linearer Triebkraftansätze, die effiziente, implizite Lösung der Energiebilanz und diverse Konzepte zur Krustenbildung im zweiten Trocknungsabschnitt. Zudem wird ein neuer Ansatz zur Berechnung von Oberflächenspannungskräften auf Basis paarweiser Kräfte vorgestellt. Durch Anpassung von Modellparametern lassen sich sowohl die Ausbildung dichter, kompakter Strukturen als auch von Hohlgranalien mit unterschiedlicher Mikroporosität simulieren. Diese Parameter können in physikalischer Weise als Bindergehalt interpretiert werden. Abschließend wird ein neuer Ansatz von SPH-Gitter-Kopplung vorgestellt, der Diffusion in der Gasphase direkt über ein unterlegtes Gitter abbildet. Berechnungen zeigen das Zurückweichen von Wasser innerhalb einer porösen Struktur infolge von Trocknung.

Acknowledgements

I am most thankful to Prof.. Nieken, who supported me through all the time. I very much enjoyed our intense scientific discussions. Also, I really valued your confidence and commitment towards me to work on a new, very different topic, when funding of my first industry driven research project became impossible during an economic crisis. You always trusted your staff and gave them free rein to approach research topics on their own. Moreover, I appreciate your understanding when it came to balancing family matters and research. The first conference on drying I went to was hosted by Prof.. Tsotsas in Magdeburg. I am most grateful and honoured that he took over as second assessor of my thesis. I also would like to acknowledge the financial support by the Deutsche Forschungsgemeinschaft, which made this work possible. Furthermore, I would like to thank Prof.. Eigenberger for valuable discussions and priceless support when it came to subsequent project applications. It has been an honour to work with you.

This thesis originates from my time as a research assistent at the Institut für Chemische Verfahrenstechnik. I am thankful to many people there and like to mention a few. Stefan Dwenger awoke my interest in doing a doctorate at the department and introduced me there. With meshfree methods being uncommon in the field of process engineering, I was glad that Franz Keller performed the pioneering work and surely endured most of the blood, sweat and tears when introducing a new method. You have always been a valuable guide. I owe many thanks to all the staff there for their support - also concerning various side projects which working as a research assistent brings along. On behalf of them I would like to thank Katrin Hungerbühler for having my back on organisational and other project issues.

Special thanks go to my long term room mates with whom I shared pain and gain. Christian Löw received me at the department and was the first I could ask questions. Matthias Rink and I shared a spiritual kinship and I am very happy about and grateful for our ongoing friendship. Manuel Hopp-Hirschler, you showed an infectious enthusiasm for research and were a great sparring partner for discussions on meshfree methods. And finally, Sebastian Gast gave me the joy of seeing a former student decide going into research. Thanks a lot for taking care of the coffee machine and your abundant supply of fair trade coffee.

Every cloud has its silver lining, and if you're stuck in research, it is the "Ska-trunde" after lunch. Matthias, Manuel and Philipp, thanks a lot for frequently saving my day. I really miss playing skat with you.

Not everything was bright during that time. Cancer and chemotherapy are nothing which I would like to repeat. Yet, it was also a time during which I was touched by the sympathy of all my colleagues. I still remember Stefan's and Matthias' visit to our home, bringing a huge pile of chocolate bars from the ICVT.

Last, but surely not least I want to thank my family. My parents gave me a loving home and supported my development from early on. Before going to school, I learnt calculating on my father's lap and found that dealing with numbers was actually fun. Later, I was equipped with various science and ex-periment kits and, of course, supported when taking up technical studies. Sadly, you were not able to share the finalisation of my thesis anymore. A big thank you to both of you!

I am deeply grateful to my wife Sarah. "For better or worse" is more than a phrase, I am fortunate that you bring it to life. Thank you for your ongoing love, for being positive, constructive and supporting! And finally, Lukas and Oliver, I am most happy to have such cool sons! Thank you for bringing laughter, joy and surprise into my life. This work is dedicated to you.

Contents

List of Symbols

Symbols

Symbol	Description	Unit
$[*]$	concentration of species $*$	$\mathrm{mol/m^3}$
$[R_{tot}]$	concentration of all polymer radical chains	$\mathrm{mol/m^3}$
α	heat transfer coefficient	$\mathrm{W/(m^2\,K)}$
β	mass transfer coefficient	$\mathrm{m/s}$
δ	Dirac delta	$\mathrm{1/m^d}$
δ	Kronecker delta	$-$
η	dynamic viscosity	$\mathrm{Pa\,s}$
γ	activity coefficient	$-$
λ_k	k^{th} moment of polymer radical chain length distribution	$\mathrm{mol/m^3}$
λ	thermal conductivity	$\mathrm{W/(m\,K)}$
μ	chemical potential	$\mathrm{J/mol}$
μ_k	k^{th} moment of both living and dead polymer chain length distribution	$\mathrm{mol/m^3}$
ν	kinetic chain length / average length of living chains	$-$
v_j^m, v_j^N	mass or molar specific volume of species j	$\mathrm{m/kg}$, $\mathrm{m/mol}$
ν_{ij}	stoichiometric coefficient of component j in reaction i	$-$

Symbol	Description	Unit
Ω	mass flux across interface to the drying gas	$kg/(m^2\,s)$
Ω^N	molar flux across interface to the drying gas	$mol/(m^2\,s)$
Φ	molecular flux of physical entity Ψ	$*$
φ	volume fraction	$-$
Ψ	arbitrary physical entity	$*$
ρ	density	kg/m^3
σ_Ψ	source of physical entity Ψ	$*$
τ^P	penalty coefficient	$-$
ξ	dimensionless coordinate	$-$
ζ_k	k^{th} moment of dead / terminated polymer chain length distribution	mol/m^3
$\Re$	universal gas constant	$J/(mol\,K)$
$\vec{a}$	acceleration	m^2/s
A	area	m^2
c	concentration	mol/m^3
c_M^{I63}	monomer concentration, at which initiator efficiency is reduced to 63 %	mol/m^3
c_p	specific heat capacity	$J/(kg\,K)$
c	colour function in CSF approach	$*$
C^τ	penalty parameter in ISPH free surface condition	$-$
C^P	parameter for proportional contribution in ISPH density correction	$-$
$\vec{d}$	driving force in MS diffusion	$1/m$
d	nr. of spatial dimensions in SPH calculation	$-$
$Đ_X$	dispersity index	$-$
D_{log10}^{lim}	parameter in artificial diffusion coefficient equation	$-$
D	mass diffusivity / Fickian diff. coefficient	m^2/s
$Đ_{jk}$	binary diffusion coefficient between species j and k in mixture	m^2/s
DP^{inst}	degree of polymerisation of polymer produced at current instant of time	$-$
$\vec{e}_i$	unit vector in spatial direction i	m
$\vec{F}$	force	N

Symbol	Description	Unit
f^{bulk}	continuous function evaluating whether a particle belongs to the particle bulk	–
f^{surf}	continuous function evaluating whether a particle belongs to the surface	–
f_d	initiator efficiency	–
g^E	Gibbs free energy	J/mol
h	mass specific enthalpy	J/kg
$\Delta h_{v,j}$	heat of evaporation of species j	J/kg
h	height	m
h	SPH kernel function's smoothing length	m
$\vec{j}$	diffusive mass flux	$\text{kg}/(\text{m}^2\,\text{s})$
$\vec{J}$	diffusive molar flux	$\text{mol}/(\text{m}^2\,\text{s})$
k_i	interaction strength of particle i in pairwise surface lateral forces	–
l_0	particle spacing	m
L	characteristic length	m
m	mass	kg
m^{evap}	evaporation related mass of SPH particle	kg
MW	molar weight	kg/mol
N	amount of substance	mol
$\vec{n}$	surface/interface normal	m
n	particle number density	$1/\text{m}^d$
p	pressure	Pa
p_v	vapour pressure	Pa
P_n	number average of polymer chain length distribution	–
P_w	mass average of polymer chain length distribution	–
$\vec{q}$	heat flux density	$\text{J}/(\text{m}^2\,\text{s})$
r	reaction rate	$\text{mol}/(\text{m}^3\,\text{s})$
r_j^F	rate of formation of component j	$\text{mol}/(\text{m}^3\,\text{s})$
r	radial coordinate	m
R	droplet radius	m
s_{ij}	interaction parameter for pairwise forces between particles i and j	N/m^d

Symbol	Description	Unit
t	time	s
T	temperature	K or °C
u	inner energy	J/kg
$\vec{v}$	velocity	m/s
V	volume	m^3
w	mass fraction	—
w_P^{crit}	parameter in artificial diffusion coefficient equation	—
W	kernel function	m^{-d}
x	mole fraction	—
$\vec{x}$	position in space	m
X_j	conversion of species j	—
Y_j	yield of species j	—

Superscripts

Symbol	Description
$*$	intermediate value
α, β	spatial dimension
0	reference point or pure substance
∞	at the surrounding
$0D, 1D$	lumped or one-dimensional model
Γ	phase boundary
att	attractive
D	diffusion related
G	gas
$inst$	at current instant of time
L	liquid
$merge$	concerning rigid body merging
N	molar averaged
$prepoly$	partial polymerisation before atomisation
R	reaction related
rep	repulsive
$surf$	surface

Subscripts

Symbol	Description
$h*$	at certain height
i,j,k	component/species in mixture
i	ID of particle of interest
j	ID of neighbouring particle
lim	limiting/threshold value
s,t	chain length / number of units in polymer chain

Species and Reactions in Spray Polymerisation

Symbol	Description
I	initiator
I^*	initiator radical
IC	consumed initiator
M	monomer
P_s	dead / terminated polymer chain of length s
R_s	chain radical / living polymer chain of length s
S	solvent
d	decomposition of initiator
i	initiation of polymer chains
p	propagation reaction
t,tc,td	termination reaction, in general, by recombination, by disproportionation
tr,trm,trs	transfer reaction, in general, to monomer, to solvent

Acronyms

Acronym	Description
BDF	backward differentiation formula
CLD	chain length distribution
CSF	continuum surface force
DAE	differential-algebraic equation
DEM	discrete element method
DoE	design of experiment
EOS	equation of state
FDM	finite differences method
FRP	free radical polymerisation
FVM	finite volume method
GP	Gaussian process
ISPH	incompressible SPH
IVP	initial value problem
LDF	linear driving force
MLS	moving least squares
MoM	method of moments
ODE	ordinary differential equation
PPE	pressure Poisson equation
QSSA	quasi-steady-state assumption
SPH	smoothed particle hydrodynamics
TVD	total variations diminishing
UNIFAC	universal quasichemical functional group activity coefficients
WCSPH	weakly compressible SPH

List of Figures

List of Tables

1. Introduction: Spray Drying and Reactive Drying Processes

Spray drying is a basic, common operation in the process industry in order to obtain powder from a liquid feed. It is applied in food and pharmaceutical processes as well as in the chemical industry. The feed - a solution, a slurry or an emulsion - is atomised into hot gas. Due to the intensive heat and mass transfer, drying takes place within seconds. The final product properties depend on many parameters like process conditions, operation mode and properties of the feed material. Production of tailored products demands extensive experiments. The question of predictively modelling the product properties has been addressed in different ways. Current state of the art are single droplet models, which typically assume spherical symmetry of the droplet and - at least in sections - a quasi-homogeneous material. Detailed simulation of structure generating processes is however not possible with these approaches. Whereas sole drying is widely applied, reactive spray drying processes are rare. Spray polymerisation has been discussed to some extend in the literature, but can still be considered as a "promising process", which offers desired properties in theory, but lacks practical applications and a theoretical model.

Both questions, the matter of structure evolution within a droplet and modelling and process evaluation of spray polymerisation will be adressed within this work by single droplet models. In the first part, droplet polymerisation will be evaluated by an extension of established, one-dimensional single droplet models of solution drying to a reactive drying model. The second part will consider structure evolution within the drop by a novel approach based on the meshfree SPH method. Focus of this work lies in the derivation and application of these new models. Especially the development of the SPH drying model concerns many implementational aspects in order to achieve the desired applicability. Drying applications will therefore not be treated in detail and the reader is referred to the respective literature (e.g. Mujumdar 2007).

1.1 Spray Polymerisation

The basic idea of spray polymerisation utilises the combination of several process steps within one apparatus. This appears attractive for several reasons. Concerning process intensification and an efficient use of resources, the reduction of process steps is favourable, especially, when small spherical polymer particles need to be obtained. Assuming that chemical reactions take place during evaporation of the solvent (Biedasek 2009, p. 15), the combination of an energy demanding (drying) process with the heat release of chemical reactions is appealing. Moreover, batch processes demand a high solvent content in order to reduce the viscosity. Heat release due to chemical reactions needs to be dissipated efficiently, whereas a small droplet provides very intensive heat and mass transfer with the surrounding gas. Despite these attractive features and a series of patents on this process (for an extensive study see Krüger 2004), these considerations have mostly been theoretical. Only very fast reactions can be carried out within the gas because of the little residence time. Especially acrylic acid and its salts appear attractive as possible applications, which has been considered experimentally on a lab scale by Franke, Moritz, and Pauer (2017). The process control within a droplet is limited to the adaption of feed composition and adjustment of drying gas properties, which may, however vary significantly throughout a dryer. A model, which takes all relevant chemical reactions into acount and predicts product properties, is therefore highly desirable. So far, Walag (2011) provides the only mathematical model of this process. By assuming the droplet as ideally mixed, he derives equations for emulsion polymerisation within a drop. A spatially resolved description, which takes inhomogeneities and transport processes within the droplet into account, is lacking yet.

1.2 Single Droplet Models for Spray Drying

Mezhericher, Levy, and Borde (2010) distinguish between three basic concepts of droplet drying models. Semi-empirical approaches are based on a characteristic drying curve, with process prediction being largely based on experimental experience. The reaction engineering approach does not take the complete processes within a droplet into account, but only solves the equations for the droplet as a whole. Its efficiency makes it interesting for the incorporation in large scale simulations of a complete spray dryer, in which more resolved models would be too costly. If the model is assumed in a lumped approximation, the model must be fitted well to experiments or previously been tuned by the third class of single droplet models, deterministic drying models. These models take all relevant processes within the drop into account and typically solve the underlying equations on a one-dimensional scale in radial direction. Azimuthal or polar effects are typically neglected under the assumption of spherical symmetry. The reactive drying model, which is derived throughout this work, falls into the last category.

Nesic and Vodnik (1991) provide a principle model of the different process periods during drying, from initial heating, to a quasi-steady-state in which heat transfer is directly converted into evaporation up to the formation of a crust, which hinders further evaporation and provokes heat up, followed by boiling and final drying. The process of crust formation has been addressed in several ways like treatment of the solid phase by population balance models (Seydel 2005; Handscomb, Kraft, and Bayly 2009) or by representation of a crust with approximate, regular (pore)geometries (Mezhericher, Levy, and Borde 2009). If solution drying is involved, the transport of a solute in radial direction is modelled (Sloth et al. 2006; Czaputa and Brenn 2012).

Concerning reactive drying processes, in particular spray polymerisation, a lumped approximation of the droplet may fall short, as spatially inhomogeneous educt concentrations change the local reaction conditions and hence polymer properties. The spray polymerisation model will therefore follow the one-dimensional models of solution drying and enhance this approach with special regard to polymerisation reactions. A different approach will be laid out concerning structure evolution. With the long term aim of predictively modelling how a structure is formed on a detailed scale, a two-dimensional model will be employed, which accounts for all underlying physical effects.

1.3 Meshfree Methods and Simulation of Structure Evolution

Meshfree methods discretise the continuum by interpolation points, so-called particles, which are moving according to a Lagrangian point of view. The term particle is not to be mistaken as a solid granular mass, aside from special approaches like the Discrete Element Method (Cundall 1971). Particles mean truly interpolation points, which are employed in order to discretise continuum equations. In this sense, implementation of physical effects can be undertaken based on first principles. Particles may bear additional properties or variables. Therefore, these methods are not bound to the equations of motion, but can be used for the solution of additional physical effects by using the meshfree discretisation operators.

The Lagrangian nature of meshfree approaches offers natural advantages over grid-based methods. When evolving interfaces are modelled by grid-based methods, the mesh has to adapted continuously and interfaces need to be tracked by approaches like Level Set (Sethian 1999) or Volume of Fluid (Hirt and Nichols 1981). This may become very costly, especially when a large number of interfaces is to be tracked and strong deformations occur. Due to particle motion, meshfree methods are self-adapting. Particle classes of various kind can be employed in order to represent different physical phases. Phase boundaries are therefore intrinsically represented by co-occurence of particles of different nature. Interface tracking is in the best case done automatically due to particle motion. The evolution of material bridges, break-up and merging can also be treated in a natural way. On the downside, there is little standard simulation software for meshfree methods, which also involves a lack of standard workflows and an increased user effort for pre- and post-processing. Furthermore, the advantage of a flexible particle distribution needs to be paid by a high computational load, as discretisation operators typically involve a much larger number of neighbouring points than in grid-based methods. Besides, meshfree methods partly lack a mathematical foundation, when an analysis can only be conducted for regular particle alignment, whereas particles are allowed to be distributed irregularly throughout a simulation. Hence, such approaches can be considerd as special methods suitable for special problems, which cannot compete with highly developed mesh-based methods in their typical applications. For non-standard cases, in which established methods are challenged, meshfree methods provide an alternative. Structure evolution can be considered as one of these particular problems.

Throughout this work, the Smoothed Particle Hydrodynamcs method (SPH) will be used, which can be considered as the first completely meshfree method in computational fluid dynamics. SPH was developed at the same time independently by Lucy (1977) and Gingold and Monaghan (1977) for the simulation of astrophyical problems. In this field it became one of the standard approaches within the next decades. In the 1990s, SPH was extended to problems concerning incompressible liquids (Monaghan 1994; Cummins and Rudman 1999), which made it interesting for engineering problems as well. Reviews of the method have been provided by Monaghan (1992, 2005, 2012), one of its original inventors. SPH applications are often found in problems involving sloshing or violent flows, which for instance concerns coastal engineering. The method proved to be suitable to calculation of large material deformations like the impact of projectiles (Stellingwerf and Wingate 1994) and for calculating the evolution of cracks (Das and Cleary 2013). Its capability of treating problems of structure evolution was shown by Keller (2015), who modelled the generation of a porous structure, determined by a considerable number of interacting physical effects.

1.3.1 Previous Applications of SPH to Drying

Drying models employing SPH are very rare, but the little number of contributions is indeed concerned with the matter of structure evolution, which underlines the special ability of SPH in this respect. Ito and Yukawa (2012) modelled the evolution of cracks within a flat drying paste by SPH. Due to the drying process, the mechanical stress inside the paste increases. If, locally, the stress exceeds the yield stress, the material will be damaged there, resulting in the initiation of a crack. Subsequently, this fracture grows further and takes up the mechanical stress within its surrounding. The process of drying itself has not been taken into account directly. Ito and Yukawa rather considered a constant increase of mechanical stress over time as a result of drying without accounting for heat and mass transfer. The model therefore provides an insight into the fracturing of a drying material, but crack initiation and growth in this approach could be linked to any phyiscal effect with continuously increasing mechanical stress and not just drying in particular.

Karunasena, Senadeera, Gu, et al. (2014) and Karunasena, Senadeera, Brown, et al. (2014a,b) examined the drying of plant cells and studied the impact of the drying process on the deformation of single cells as well as on cell clusters. The morphological behaviour was captured by an SPH-DEM hybrid model, in which the cell walls are respresented as linked DEM particles and the cell interior is modelled as a highly viscous Newtonian liquid using SPH. The drying kinetics rely on the differences in osmotic pressure over the cell walls with the mass loss due to drying being dedicated to all SPH particles within a cell in equal measure. With the mass of the particles being diminished, their volume is reduced as well, resulting in cell shrinkage. The structural behaviour is determined by the parameterisation of several attractive and repulsive forces between DEM particles and according to their interactions with SPH particles.

Detailed, spatially resolved simulations of drying kinetics have not been conducted in SPH so far.

2. THEORETICAL PRINCIPLES

In the following, the underlying principles being fundamental for the derivation of drying models will be laid out. Typically, one would be short on this topic when presenting a classic single droplet drying model. However, this work is also concerned with an appropriate and sound representation of the polymer phase and a consistent derivation of balance equations. Indeed, there are still drying models being published, in which transport equations are not mass-conservative. Such approaches are neither sufficient for more complicated applications like spray polymerisation nor are the underlying simplifications reasonable in general. Therefore, the principles of diffusion and reaction driven convection and their consideration in transport equations will be discussed to a longer extend.

2.1 Transport Equations

Transport of an entity can be modelled mathematically via a partial differential equation in differential or integral form. In the following only differential formulations will be used. In general, transport of an arbitrary entity ψ can be expressed as

$$\frac{\partial \psi}{\partial t} + \nabla \left(\psi \vec{v^*} \right) + \nabla \phi_\psi = \sigma_\psi. \tag{2.1}$$

The temporal change of the respective entity, $\frac{\partial \psi}{\partial t}$, is called the accumulation term. The second term denotes convective transport with the continuum velocity v^*, whereas ϕ_ψ indicates the molecular flux of the property ψ (Bird, Stewart, and Lightfoot 2002, p. 588). Sources and sinks are denoted as σ_ψ. Table 2.1 gives an overview of the most common properties with their respective fluxes and sinks/sources. $\vec{v^*}$ is an average fluid velocity, where the base for averaging can in general be arbitrarily chosen (Taylor and Krishna 1993, p. 3ff). Most commonly

Table 2.1: Transport properties.

equation	property ψ	velocity v^*	flux ϕ_ψ	sources σ_ψ
continuity eq.	ρ	$\vec{v}$		
momentum eq.	$\rho\vec{v}$	$\vec{v}$	$-\Pi$	$\rho\vec{f}$
mass bal. species j	ρ_j	$\vec{v}$	$\vec{j}_j$	$r_j^F MW_j$
energy balance	ρu	$\vec{v}$	$\vec{q}+\sum_j \vec{j}_j h_j + p\vec{v}$	$-\tau:\nabla\vec{v}+\vec{v}\cdot\nabla p$
continuity eq. molar	c	$\vec{v}^N$		$\sum_j r_j^F$
molar bal. species j	c_j	$\vec{v}^N$	J_j^N	r_j^F

the mass average velocity $\vec{v}$ and the molar average velocity $\vec{v}^N$ are employed for modelling transport of mass based properties such as (partial) densities and mass fractions or molar values as concentrations and mole fractions, respectively.

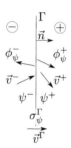

Figure 2.1: Transport across a phase boundary Γ.

Equation 2.1 is the conservation equation of a property in a continuum, where a smooth distribution of the respective values can be assumed. In contrast, a property may change discontiniously at a phase transition, with an abrupt change in physical behaviour. Mathematically the interface between two phases – and + is assumed to be infinitesimally thin without storage capacity, at which the respective properties may change jumpwise between both phases (see Figure 2.1). In- and outgoing fluxes and sinks/sources need to add up to zero. A general balance equation at an interface Γ is (comp. Taylor and Krishna 1993, p. 9f)

$$\left(\psi^+\vec{v}^+ + \phi_\psi^+ - \psi^+\vec{v}^\Gamma - \psi^-\vec{v}^- - \phi_\psi^- + \psi^-\vec{v}^\Gamma\right)\cdot\vec{n} - \sigma_\psi^\Gamma = 0. \qquad (2.2)$$

Superscripts – and + denote the phases on both sides of the interface, where the interface normal $\vec{n}$ is directing from – to +. Additionally to convective and diffusive transport the motion of the phase boundary Γ itself with the interface velocity $\vec{v}^\Gamma$ needs to be taken into account. Equation 2.2 is the basis for the derivation of boundary conditions in mathematical models. These will be considered in detail, when the equations for reactive drying models of single droplets are derived in chapter 3.

2.1.1 Transport in a Mass Averaged System

Mass and Momentum Balance Equations

Due to the basic principles of mass conservation and the connection of momentum with the mass or density of a fluid element, it is straightforward to model the laws of hydrodynamics in a mass averaged notation with the velocity $\vec{v}$. $\psi = \rho$ yields the mass balance or continuity equation

$$\frac{\partial \rho}{\partial t} = -\nabla (\rho \vec{v}). \tag{2.3}$$

With $\psi = \rho \vec{v}$ the momentum balance (Navier-Stokes equations) in conservative form can be obtained to

$$\frac{\partial \rho \vec{v}}{\partial t} = -\nabla (\rho \vec{v} \vec{v} - \Pi) + \rho \vec{f}, \tag{2.4}$$

in which Π denotes the Cauchy stress tensor and $\vec{f}$ is a vector of external forces (as force per unit mass - in unit of an acceleration) such as gravity and surface tension. Applying the product rule on equation 2.4 and subtracting the continuity equation the non-conservative momentum balance can be obtained (Ferziger and Peric 2007, p. 10)

$$\rho \frac{\partial \vec{v}}{\partial t} = -\rho \vec{v} \nabla \vec{v} + \nabla \Pi + \rho \vec{f}. \tag{2.5}$$

Whereas in simulations both formulations yield similar results on very fine meshes, for coarse discretisations the non-conservative form is prone to additional errors (Ferziger and Peric 2007, p. 10). The Cauchy stress tensor is usually split into a diagonal tensor corresponding to the fluid pressure p and the deviatoric stress tensor τ

$$\Pi = -pI + \tau \tag{2.6}$$

$$\frac{\partial \vec{v}}{\partial t} = -\vec{v} \nabla \vec{v} - \frac{\nabla p}{\rho} + \frac{\nabla \tau}{\rho} + \vec{f}. \tag{2.7}$$

The Navier-Stokes equations and the mass balance provide a set of $d + 1$ equations (d being the spatial dimensionality of the problem), but contain $d + 2$ unknowns - the density, d components of the velocity vector and the pressure. In compressible flows the pressure is dependent on the density and can be evaluated by an equation of state (EOS). Fluid density and pressure are not coupled in case of incompressible liquids ($p \neq f(\rho)$) and, without further considerations, the system is underdetermined. The most common assumption for incompressible flows is that the density does not change at all. With $\rho = const.$, the velocity field has to be divergence-free, as the continuity equation reduces to

$$\nabla \vec{v} = 0. \tag{2.8}$$

Applying this constraint to the Navier-Stokes equations, the divergence of the velocity change over time has to be zero as well. Taking the divergence of the momentum balance 2.7 provides a Poisson equation for the unknown pressure

$$\nabla \left(\frac{\nabla p}{\rho} \right) = \frac{1}{\rho} \nabla^2 p = \nabla \left(-\vec{v} \nabla \vec{v} + \frac{\nabla \tau}{\rho} + \vec{f} \right). \tag{2.9}$$

The detailed formulation of this pressure Poisson equation (PPE) and its implementation is depending on the underlying numerical algorithm. An alternative to this procedure is the assumption of a slight compressibility, so that $p = f(\rho)$. In this case the relation between pressure and density typically is established by a stiff, artificial equation of state. Both approaches will be discussed more explicitly for the Smoothed Particle Hydrodynamics method in sections 5.3 and 5.4.

The stress tensor is formulated according the underlying rheology. In case of Newtonian liquids, τ is expressed as (Ferziger and Peric 2007, p. 6)

$$\tau = \eta \left(\nabla \otimes \vec{v} + (\nabla \otimes \vec{v})^T + \frac{2}{3} \nabla \vec{v} \right) \stackrel{\rho=const.}{=} \eta \left(\nabla \otimes \vec{v} + (\nabla \otimes \vec{v})^T \right). \tag{2.10}$$

Mass Balance Equations of Single Components

Transport equations for single components j can be derived by $\psi = \rho_j$ to

$$\frac{\partial \rho_j}{\partial t} = -\nabla \left(\rho_j \vec{v} + \vec{j}_j \right) + r_j^F MW_j. \tag{2.11}$$

$\vec{j}_j$ indicates the diffusive flux. The source term accounts for production or consumption of the respective component by chemical reactions with MW_j being its molar weight. The rate of formation r_j^F denotes the molar conversion of j with respect to time and is determined by the reaction rates r_i of all chemical reactions i and the corresponding stoichiometric coefficients v_{ij}

$$r_j^F = \sum_i v_{ij} r_i. \tag{2.12}$$

The continuity equation 2.3 evolves from the addition of all component transport equations as well. In mass based notation, the sum of all source terms is zero, as mass is neither produced nor destroyed by chemical reactions (whereas the number of moles may change). Additionally, the sum of all diffusive fluxes adds up to zero. Using the product rule on equation 2.11 and subtracting the continuity equation, transport of a component with respect to its mass fraction $w_j = \rho_j / \rho$ can be expressed as follows:

$$\frac{\partial w_j}{\partial t} = -\vec{v} \nabla w_j + \frac{-\nabla \vec{j}_j}{\rho} + r_j^F \frac{MW_j}{\rho}. \tag{2.13}$$

The total fluid density ρ can be calculated from the mass fractions using the respective densities of the pure substances ρ_j^0

$$\rho = \frac{m}{\sum V_j} = \frac{m}{\sum \frac{m_j}{\rho_j^0}} = \frac{1}{\sum \frac{w_j}{\rho_j^0}}. \tag{2.14}$$

However, this relation requires the premise, that the volumes of the single components V_j can be calculated independently and add up to the total volume without any excess volumes. The total mass flux $\vec{m}_j$ of a component j is

$$\vec{m}_j = \rho_j \vec{v} + \vec{j}_j. \tag{2.15}$$

Transport of Energy

The balance of inner energy ($\psi = \rho u$)

$$\frac{\partial (\rho u)}{\partial t} = -\nabla \left(\rho u \vec{v} + + p \vec{v} + \vec{q} + \sum \vec{j}_j h_j \right) - \tau : \nabla \vec{v} + \vec{v} \nabla p. \tag{2.16}$$

can be converted into enthalpy notation using $\rho u = \rho h - p$

$$\frac{\partial (\rho h)}{\partial t} = -\nabla \left(\rho h \vec{v} + \vec{q} + \sum \vec{j}_j h_j \right) - \tau : \nabla \vec{v} + \vec{v} \nabla p + \frac{\partial p}{\partial t}. \tag{2.17}$$

h and h_j are the specific enthalpies of the mixture and the single components, respectively. The molecular flux of enthalpy h involves both heat conduction with the heat flux $\vec{q}$ as well as enthalpy transport due to molecular fluxes of the single components j. The source terms consist of an irreversible part in consequence of viscous dissipation and a reversible contribution due to changes in fluid pressure. In case of moderately viscous media the sources can usually be neglected. In most physically relevant cases the heat flux $\vec{q}$ is calculated by Fourier's first law

$$\vec{q} = -\lambda \nabla T. \tag{2.18}$$

λ is the thermal conductivity and T the temperature. The average enthalpy h is

$$h = \sum w_j h_j \tag{2.19}$$

$$h_j = h_j^0 + \int_{T^0}^{T} c_{p,j} dT, \tag{2.20}$$

in which the enthalpy of a component j is defined with respect to a standard enthalpy h_j^0 at a reference temperature T^0. In the general case of non-constant heat capacities, evalutation of the enthalpy according to equation 2.17 and solution of the algebraic constraints 2.19 and 2.20 provide the temperature. When heat capacities c_{p_j} are practically constant and neglecting the mechanical source terms, a temperature balance can directly be obtained:

$$h_j = h_j^0 + c_{p,j} \left(T - T^0 \right) \tag{2.21}$$

$$c_p = \bar{c}_p = w_j c_{p,j} \tag{2.22}$$

$$h = \sum w_j h_j^0 + c_p \left(T - T^0 \right) \tag{2.23}$$

$$\frac{\partial T}{\partial t} = -\vec{v} \nabla T - \frac{\sum c_{p\,j} \vec{j}_j}{\sum \rho_j c_{p_j}} \nabla T - \frac{\nabla \vec{q}}{\sum \rho_j c_{p_j}} - \frac{\sum r_j^F MW_j h_j}{\sum \rho_j c_{p_j}}. \tag{2.24}$$

The reaction term respective the rates of formation of all components can be expressed by the heat of reaction $\Delta h_{R,i}$ of all chemical reactions i, as well, using equation 2.12 and the definition of the molar based specific enthalpy $h_j^N = MW_j h_j$

$$\Delta h_{R,i} = \sum_j v_{ij} h_j^N \tag{2.25}$$

$$\sum_j r_j^F MW_j h_j = \sum_i r_i \Delta h_{R,i}. \tag{2.26}$$

The Lewis number denotes the ratio between mass diffusivity D and thermal diffusivity

$$Le = \frac{\rho D c_p}{\lambda}. \tag{2.27}$$

In case of a small Lewis number, heat transport due to mass diffusion (second term in equation 2.24) is therefore subordinate compared to thermal conduction (third term).

2.1.2 Transport in a Molar Averaged System

According to Bird, Stewart, and Lightfoot (2002, p. 584) the continuity equation and transport of the single components in a molar system

$$\frac{\partial c}{\partial t} = -\nabla \left(c \vec{v}^N \right) + \sum_j r_j^F \tag{2.28}$$

$$\frac{\partial c_j}{\partial t} = -\nabla \left(c_j \vec{v}^N + \vec{J}_j^N \right) + r_j^F, \tag{2.29}$$

are an equivalent expression to equations 2.3 and 2.11. $\vec{J}_j^N$ is the molecular molar flux of the component j with respect to the molar average velocity v^N. Again, the continuity equation is the combination of the single component transport equations. However, whereas in a mass averaged system the reaction term vanishes in the continuity equation 2.3, it is still present in a molar based notation, as chemical reactions preserve mass but are not cardinally conservative concerning the number of moles. The component balance equation with respect to the mole fraction x_j is

$$\frac{\partial x_j}{\partial t} = -\vec{v}^N \nabla x_j + \frac{-\nabla \vec{J}_j^N}{c} + \frac{r_j^F}{c}. \tag{2.30}$$

2.1.3 Reference Velocities and Conversion between Systems

The velocity of a component $\vec{v}_j$ denotes the motion of this component with respect to a fixed coordinate reference frame (Bird, Stewart, and Lightfoot 2002, 535f; Taylor and Krishna 1993, p. 3), independently from an average velocity $\vec{v}^a$. An arbitrarily chosen averaging can be undertaken by

$$\vec{v}^a = \sum a_j \vec{v}_j, \tag{2.31}$$

where the weighting factors a_j satisfy $\sum a_j = 1$. The mass and molar average velocities are calculated according to

$$\vec{v} = \sum_j w_j \vec{v}_j \tag{2.32}$$

$$\vec{v}^N = \sum_j x_j \vec{v}_j. \tag{2.33}$$

A component's velocity $\vec{v}_j$ can be calculated from mass or molar fluxes by (Bird, Stewart, and Lightfoot 2002, p. 537)

$$\vec{v}_j = \vec{v} + \frac{\vec{j}_j}{\rho_j} = \vec{v}^N + \frac{\vec{J}_j^N}{c_j}. \tag{2.34}$$

Inserting this equation into the definitions of mass and molar average velocities the following conversion formulae can be derived

$$\vec{v} = \vec{v}^N + \frac{1}{\rho} \sum_k \vec{J}_k^N MW_k \tag{2.35}$$

$$\vec{v}^N = \vec{v} + \frac{1}{c} \sum_k \frac{\vec{j}_k}{MW_k}. \tag{2.36}$$

Conversion of Diffusive Fluxes

Diffusive fluxes are defined as the difference between the velocity of a component and the average velocity

$$\vec{j}_j = \rho_j \left(\vec{v}_j - \vec{v} \right) \tag{2.37}$$

$$\vec{J}_j^N = c_j \left(\vec{v}_j - \vec{v}^N \right). \tag{2.38}$$

In any system the sum of all molecular fluxes adds up to zero

$$\sum_j \vec{j}_j = 0, \qquad \sum_j \vec{J}_j^N = 0. \tag{2.39}$$

By reason of this rule a direct conversion between molar and mass averaged fluxes via multiplication/division with the molar mass is not generally valid. In fact, by doing this one obtains diffusive mass fluxes relative to the molar average velocity and vice versa. These fluxes do not necessarily add up to zero and are of minor relevance in mathematical models. Rather, with (Bird, Stewart, and Lightfoot 2002, p. 537)

$$\sum_j \frac{\vec{j}_j}{MW_j} = c \left(\vec{v}^N - \vec{v} \right) \tag{2.40}$$

$$\sum_j \vec{J}_j^N MW_j = \rho \left(\vec{v} - \vec{v}^N \right) \tag{2.41}$$

and by inserting equation 2.34 the conversion laws

$$\vec{j}_j = \vec{J}_j^N MW_j - w_j \sum_k \vec{J}_k^N MW_k \tag{2.42}$$

$$\vec{J}_j^N = \frac{\vec{j}_j}{MW_j} - x_j \sum_k \frac{\vec{j}_k}{MW_k} \tag{2.43}$$

can be derived. Hence, a conversion of a flux between both systems by simple multiplication with the molar mass is only applicable in case of (nearly) identical molar weights of all species in the mixture or when the respective component is highly diluted. Somewhat more complicated conversion laws based on matrix multiplications can be found at Taylor and Krishna (1993, 6f). The conversion of diffusive fluxes from molar notation to a mass based one is more relevant in practice, as diffusion is a molecular phenomenon depending on the chemical potential, which is naturally described using molar values.

Conversion of Component Balance Equations

With the velocity of a component $\vec{v}_j$ and equation 2.34 the components' transport equations can be formulated independently from an averaged velocity

$$\frac{\partial \rho_j}{\partial t} = -\nabla \left(\rho_j \vec{v}_j \right) + r_j^F MW_j \tag{2.44}$$

$$\frac{\partial c_j}{\partial t} = -\nabla \left(c_j \vec{v}_j \right) + r_j^F . \tag{2.45}$$

Inserting equation 2.34 yields component balance equations for partial densities in a molar averaged frame of reference and for concentrations related to mass averaged transport values, respectively

$$\frac{\partial \rho_j}{\partial t} = -\nabla \left(\rho_j \vec{v}^N + MW_j \vec{J}_j^N \right) + r_j^F MW_j \tag{2.46}$$

$$\frac{\partial c_j}{\partial t} = -\nabla \left(c_j \vec{v} + \frac{1}{MW_j} \vec{j}_j \right) + r_j^F . \tag{2.47}$$

This interrelation can be convenient, when both the concentration as well as the partial density of a species need to be modelled, for instance in order to obtain the molar mass of a polymer. In this case both values can be evaluated using only one frame of reference and not a combination of molar and mass based transport.

2.1.4 Eulerian and Lagrangian Frames of Reference

The transport equations stated out above have been derived for a fixed system of coordinates. This refers to a Eulerian point of view, which means that an observer, keeping track of the fluid movement, is spatially fixed. In computational fluid dynamics, Eulerian methods therefore employ a grid, which can be spatially fixed or experience alterations, which are not necessarily connected to the fluid movement. In contrast, a Lagrangian observer is attached to a fluid element and moves along the lines of flow. The motion of discretisation points in a Lagrangian method is therefore directly bound to fluid movement. Because these approaches keep track of small mass elements in the fluid, often the term "'particle methods"' is used, considering these mass elements as fluid particles. Still, this expression cannot be interpreted as particles in the sense of granular masses (like billiard balls or the DEM method as a numerical approach), but rather as interpolation points without a definite shape, representing a finite fluid mass element. An Eulerian transport equation of an entity Ψ may be transformed into a Lagrangian frame of reference by the material derivative

$$\frac{D\Psi}{Dt} = \frac{\partial \Psi}{\partial t} + \vec{v}\nabla\Psi. \tag{2.48}$$

The left hand side gives the change of the respective quantity with respect to an infinitesimally small mass element (hence the terminus material derivative) according to a Lagrangian point of view. The first term on the right hand side contains the Eulerian continuum equation, whereas the second term refers to the change of a quantity induced by the moving observer.

The most important equations in Lagrangian methods are the mass and momentum balances. Applying 2.48 on equations 2.3 and 2.7 one obtains

$$\frac{D\rho}{Dt} = -\rho\nabla\vec{v} \tag{2.49}$$

$$\frac{Dv}{Dt} = -\frac{\nabla p}{\rho} - \frac{\nabla \tau}{\rho} + \vec{f} \tag{2.50}$$

for the continuity and Navier-Stokes equations in a Lagrangian frame of reference.

2.2 Diffusion

A mixture of different components as a whole only experiences convective transport with the averaged velocity $\vec{v}^*$. The motion of single components can be different due to several reasons such as concentration gradients, electric fields etc. Diffusive fluxes describe the transport of the single components in relation to the averaged velocity in the chosen frame of reference as indicated in equations 2.37 and 2.38. The most common approach for diffusion in mathematical models is Fickian diffusion, either as (pseudo) binary diffusion or as a generalised Ficks law for a multi-component mixture (Taylor and Krishna 1993, pp. 19, 52). However, a general description of diffusive behaviour in a mixture is given by the Maxwell-Stefan equations, which can be derived from thermodynamics in a consistent manner. Fick's diffusion, correctly expressed, is a special case of the Maxwell-Stefan approach for ideal thermodynamics.

2.2.1 Fickian Diffusion

Fick (1855) conducted experiments on diffusion of salt in a binary solution. He proposed an analogy of diffusive mass to diffusive heat transfer so that Fourier's second law could be transferred to problems of matter diffusion due to concentration gradients. Fick's first law for molecular mass and molar fluxes in a binary mixture is (Bird, Stewart, and Lightfoot 2002, pp. 515, 535; Taylor and Krishna 1993, pp. 50, 52)

$$\vec{j}_j = \rho_j \left(\vec{v}_j - \vec{v} \right) = -\rho D \nabla w_j \tag{2.51}$$

$$\vec{J}_j^N = c_j \left(\vec{v}_j - \vec{v}^N \right) = -c D \nabla x_j, \tag{2.52}$$

where the diffusive fluxes have been defined with respect to the mass and molar average velocity, respectively. Alternatively, one can express molecular fluxes in the volume average velocity frame (Taylor and Krishna 1993, p. 51)

$$\vec{j}_j^V = \rho_j \left(\vec{v}_j - \vec{v}^V \right) = -D \nabla \rho_j \tag{2.53}$$

$$\vec{J}_j^V = c_j \left(\vec{v}_j - \vec{v}^V \right) = -D \nabla c_j. \tag{2.54}$$

Misleadingly, these laws are sometimes applied as Fickian diffusion in mass or molar average systems (e.g. in Baehr and Stephan 2010, p. 79), taking up the analogy of diffusive mass transport with heat conduction and Fourier's first law.

However, as the sum of all diffusive fluxes has to be zero, this is only permissible for constant mixture densities or concentrations (then, the volume average velocity is equal to the mass or molar averaged one). Using the conversion laws between mass and molar diffusive fluxes 2.42 and 2.43, it can easily be shown that Fickian fluxes according to equations 2.51 and 2.52 can be transformed consistently into one another, which only applies to expressions 2.53 and 2.54 when the mixture density or concentration is constant.

In multicomponent mixtures, Fick's first law can be generalised in a matrix notation (Taylor and Krishna 1993, 53f)

$$\left(\vec{j}\right) = -\rho \left[D^0\right](\nabla w) \tag{2.55}$$

$$\left(\vec{J^N}\right) = -c\left[D\right](\nabla x), \tag{2.56}$$

in which $\left(\vec{j}\right)$ or $\left(\vec{J^N}\right)$ and (∇w) or (∇x) denote column matrices of diffusive fluxes and gradients of mass or mole fractions, respectively. The entry D_{ij} in the diffusion coefficient matrix denotes the impact on the diffusive flux of component i when species j exhibits a gradient in mass or molar fraction, respectively. Matrices of diffusion coefficients with different velocity reference - $[D]$ in molar avereraged and $[D^o]$ in mass averaged notation - are generally not identically.

A special case of multicomponent Fickian diffusion evolves assuming that the diffusive flux of each component only depends on the mass or mole fraction gradient of the respective component itself. Consequently, when $\vec{j}_i$ is independent of all $w_{j \neq i}$, the matrix of diffusion coefficients becomes diagonal with all D_{ij} being zero for $j \neq i$. As the sum of all diffusive fluxes is to be zero and additionally $\sum \nabla w_j = \sum \nabla x_j = 0$, all coefficients D_{ii} must be identical. In this simple case, there is only one single, scalar diffusion coefficient D for all components

$$\left(\vec{j}\right) = -\rho D(\nabla w) \tag{2.57}$$

$$\left(\vec{J^N}\right) = -c D(\nabla x). \tag{2.58}$$

Considering the single components these expressions are equal to equations 2.51 and 2.52 in binary diffusion. Hence, this case can be considered as a pseudo-binary Fickian diffusion. Despite this simplification to the real behaviour of mixtures, this approach can be a convenient expression to study the principle influence of diffusion with only one parameter D in a mathematical model in comparison to other physical and chemical effects, especially, when it is difficult to obtain the various D_{ij} values.

2.2.2 Maxwell-Stefan Diffusion

Whereas Fick's work was concerned with binary diffusion, in the 19th century Maxwell and Stefan independently considered the diffusion in multi-component mixtures. Diffusive transport of a species j in a mixture of n different components is aroused by a driving force $\vec{d}_j$ (Taylor and Krishna 1993, p. 19)

$$\vec{d}_j = -\sum_{k=1}^{n} \frac{x_j x_k (\vec{v}_j - \vec{v}_k)}{Đ_{jk}}. \tag{2.59}$$

Maxwell-Stefan and Fickian diffusion coefficients are identical for an ideal binary mixture. Generally, this is not the case.

With $\vec{v}_j = \frac{\vec{J}_j^N}{x_j c} + \vec{v}^N$ (equation 2.34) a relationship between driving force and diffusive fluxes can be derived

$$\vec{d}_j = \sum_{k=1}^{n} \frac{x_j \vec{J}_k^N - x_k \vec{J}_j^N}{c Đ_{jk}}. \tag{2.60}$$

The generalised driving force $\vec{d}_j$ (Taylor and Krishna 1993, S. 29)

$$c \Re T \vec{d}_j = c_j \nabla_{T,p} \mu_j + (\varphi_j - w_j) \nabla p - \rho_j \left(\vec{f}_j - \sum_k \vec{f}_k \right) \tag{2.61}$$

is affected by deviations from chemical and mechanical equilibrium as well as by external forces $\vec{f}$ (for instance concerning ions in an electric field, provided as force per unit mass). μ_j and φ_j are the chemical potential and the volume fraction of component j, respectively. In droplet drying applications, diffusive transport is dominated by gradients in the chemical potentials or, accordingly, concentration differences of the single components in the mixture. Neglecting the other contributions, equation 2.61 becomes with $\mu_j = \mu_j^0 + \Re T \ln (\gamma_j x_j)$

$$\vec{d}_j = \frac{x_j}{\Re T} \nabla_{T,p} \mu_j = \frac{x_j}{\Re T} \sum_k \frac{\partial \mu_j}{\partial x_k}\bigg|_{T,p,\mu_{l \neq k}} \nabla x_k = x_j \sum_k \frac{\partial \ln (\gamma_j x_j)}{\partial x_k}\bigg|_{T,p,\gamma_{l \neq k}} \nabla x_k$$

$$\vec{d}_j = \sum_k \left(\delta_{jk} + x_j \frac{\partial \ln \gamma_j}{\partial x_k}\bigg|_{T,p,\gamma_{l \neq k}} \right) \nabla x_k. \tag{2.62}$$

Only $n-1$ driving forces are linearly independent, as $\sum_j \vec{d}_j = 0$ (Taylor and Krishna 1993, p. 24). With the diffusive flux of component n being a linear combination of the fluxes of the other components ($\vec{J}_n^N = -\sum_{k=1}^{n-1} \vec{J}_k^N$), the n-th component commonly is eliminated from equation 2.60

$$c\vec{d}_j = \sum_{k=1;k\neq j}^{n-1} \left(\frac{x_j}{\text{Ð}_{jk}} - \frac{x_j}{\text{Ð}_{jn}} \right) \vec{J}_k^N - \left(\frac{x_j}{\text{Ð}_{jn}} + \sum_{k=1;k\neq j}^{n} \frac{x_k}{\text{Ð}_{jk}} \right) \vec{J}_j^N. \qquad (2.63)$$

Again, a matrix expression can be obtained, if all $n-1$ driving forces are merged into a column matrix $\left(\vec{d} \right)$ and, accordingly, the diffusive fluxes into the matrix $\left(\vec{J}^N \right)$

$$c\left(\vec{d} \right) = -B\left(\vec{J}^N \right), \qquad (2.64)$$

in which the matrix B is

$$B_{jj} = \sum_{k=1;k\neq j}^{n} \frac{x_k}{\text{Ð}_{jk}} + \frac{x_j}{\text{Ð}_{jn}}$$

$$B_{jk} = -x_j \left(\frac{1}{\text{Ð}_{jk}} - \frac{1}{\text{Ð}_{jn}} \right). \qquad (2.65)$$

With equation 2.62 the matrix of the driving forces can be expressed by

$$\left(\vec{d} \right) = \Gamma \left(\nabla x \right). \qquad (2.66)$$

The matrix Γ accounts for the non-ideality of thermodynamics

$$\Gamma_{jk} = \sum_{k=1}^{n-1} \delta_{jk} + x_j \left. \frac{\partial ln\gamma_j}{\partial x_k} \right|_{T,p,\gamma_{l\neq k}}. \qquad (2.67)$$

The diffusive fluxes are then obtained by

$$\left(\vec{J}^N \right) = -cB^{-1}\Gamma \left(\nabla x \right). \qquad (2.68)$$

This matrix notation applies to the diffusive fluxes of $n-1$ components and the n-th flux needs to be evaluated by the closing condition. In case of ideal thermodynamics Γ becomes the identity matrix and the generalised Fick's law 2.56 is obtained. Assuming additionally, that all binary diffusion coefficients are identical $\text{Ð}_{jk} = D$, it follows directly from equation 2.60 that

$$\left(\vec{J}^N \right) = -cD \left(\nabla x \right), \qquad (2.69)$$

which is the pseudo-binary Fickian diffusion from equation 2.58 or, in a binary system, Fick's law 2.52. If the logarithm of the activity coefficients is provided - e.g. by g^E-models - equation 2.68 can be expressed as

$$\left(\vec{J}^N \right) = -cB^{-1} \left[(\nabla x) + (x\nabla \ln \gamma) \right]. \qquad (2.70)$$

2.2.3 Determination of Diffusion Coefficients

Diffusion coefficients are the prefactors determining the rate of diffusive equilibration of spatial gradients in a mixture. Generally, they are not constant, but depending on the mixture's composition. Experiments are therefore needed in order to obtain diffusion coefficients at varying concentrations. If at least the values in infinite dilution of the different species are known, averaging between these can be used for estimating the diffusion coefficient in between for concentrated mixtures. Typically arithmetic or geometric means, weighted by the mole fractions of the involved components, are applied (Taylor and Krishna 1993, p. 76).

If no measurements are available or an extrapolation to different temperatures is needed, diffusion coefficients can be calculated based on theoretical (e.g. the kinetic gas theory) or (semi-)empirical correlations. In this work, Fuller's equation (Taylor and Krishna 1993, p. 68)

$$D_{12} = \frac{1.013 \times 10^{-2}}{p} \frac{\sqrt{\frac{(MW_1 + MW_2)}{MW_1 MW_2}}}{\left(\sqrt[3]{V_1} + \sqrt[3]{V_2}\right)} T^{1.75} \tag{2.71}$$

is used in order to estimate diffusion coefficients of volatile components in the drying gas. The molecular diffusion volumes V_1 and V_2 are summed up from atomic contributions. Molar weights have to be provided in g/mol, pressure and temperature in Pa and K, respectively, in order to obtain diffusion coefficients in m^2/s.

Diffusion in the liquid phase is calculated according to pseudo-binary Fickian diffusion with an effective diffusion coefficient acting the same way on all components in this work. Its value is set dependent on the polymer weight fraction (see also section 4.1). For an overview on various approaches concerning diffusion coeffient estimation in liquid mixtures, see e.g. Taylor and Krishna (1993, 73f).

2.3 Modelling of Free Radical Polymerisation

Polymer molecules consist of linear or branched chains, in which a large number of generic subunits (of only one or a few different kinds) are strung together to macromolecules. The chain lengths or degree of polymerisation correspond with the number of subunits within a single chain. For a polymer species as a whole this number is not constant but rather present as a chain length distribution (CLD). Polymer properties strongly depend on the nature of this distribution, which makes it necessary to model this value in reaction engineering. Typically, characteristic values like the number and weight average and dispersity are of major interest, as these can be easily obtained from experiments, rather than the exact progression of the whole distribution.

The trivial approach of setting up respective equations for polymer molecules of each possible chain length is neither applicable nor necessary in order to obtain valid predictions for averaged properties of the chain length distribution. As the reaction terms for polymer molecules of different length are very similar, the mathematical redundancy can be employed in order to derive statistically averaged reaction approaches with a manageable number of equations. A number of various methods can be found at Ray (1972). Very well-established are the quasi-steady-state assumption (QSSA), which utilises algebraic equations in order to compute statistical values solely based on the educt concentration, and the method of moments (MoM), in which differential equations for the statistical moments of the chain length distribution are derived. As quasi industry standard, the commercial software Predici is widely used, which applies an adaptive h-p-Galerkin method in order to calculate the complete chain length distribution (Budde and Wulkow 1991). In the following, only free radical polymerisation (FRP) shall be considered in detail. Other kinds of polymerisation reactions are very similar from a mathematical point of view, so that model equations can be derived in an analogous way. Furthermore, only homopolymerisation will be considered. An extension to copolymerisation is, however, straightforward as long as just additional statistical moments need to be calculated.

2.3.1 Reactions in Free Radical Polymerisation

A typical reaction scheme of free radical polymerisation is provided in Table 2.2. The system contains an initiator I, which decomposes into primary radicals $I^\bullet$. Chain initiation takes place, when these free radicals attack a monomer M and, in doing so, generate a living chain / chain radical R_1 of length one. The placeholder IC denotes consumed initiator and is of no further interest concerning polymer reactions. Continually, monomer molecules attach at the radical part of the chain and increase the chain length, at which the radical moves to the newly appended monomer unit. This chain propagation is the main reaction in

Table 2.2: Basic chemical reactions in free radical polymerisation.

reaction	mechanism		
initiator decomposition	I	$\xrightarrow{k_d}$	$2f_d I^\bullet$
chain initiation	$2M + 2I^\bullet$	$\xrightarrow{k_i}$	$IC + 2R_1$
one step initiation	$I + 2f_d M$	$\xrightarrow{k_d}$	$IC + 2f_d R_1$
propagation	$R_s + M$	$\xrightarrow{k_p}$	R_{s+1}
termination by recombination	$R_s + R_t$	$\xrightarrow{k_{tc}}$	P_{s+t}
termination by disproportionation	$R_s + R_t$	$\xrightarrow{k_{td}}$	$P_s + P_t$
transfer to monomer	$R_s + M$	$\xrightarrow{k_{trm}}$	$P_s + R_1$
transfer to polymer	$R_s + P_t$	$\xrightarrow{k_{trp}}$	$P_s + R_t$

free radical polymerisation. Termination of this process takes place, when two living chains R react and their radicals are neutralised. In this case both chains may remain as two dead chains P (disproportionation) or build one single, long dead chain (recombination). Furthermore, radical transfer to monomer or a different transfer agent and transfer to polymer can occur. Transfer of a radical to polymer reactivates a dead chain, whereas typically a side-chain is created. Other reactions such as backbiting will not be taken into account in the following. The consideration is also restricted to living chains in which the radical is located at the chain's end (primary radicals).

Most commonly in polymer reaction engineering, it is assumed that polymerisation reactions kinetics are independent from the chain lengths of the reactands. This is indeed a simplification, which is not true for very short chains, where propagation rates may be considerably faster (Hutchinson 2005, p. 158; Gridnev and Ittel 1996). Still, these variations diminish rapidly with growing chain length and the amount of such short chains is rather low. In using this assumption, mathematically uniform descriptions of the single reactions are obtained, which can be reduced efficiently to convenient formulae. Polymer reaction engineering methods relying on this assumption have proven to provide accurate descriptions of polymerisation processes. The basic reactions provided in Table 2.2 and their formation rates for the different species will be explained shortly in the following.

Initiation

Chain initiation in FRP can be induced by a variety of mechanisms such as initiator decomposition and photo initiation. One very common way is to provide a starter, which thermally decomposes into primary radicals within a reaction of first order. The rate of formation of the initiator therefore only depends on its concentration $[I]$

$$r_I^F\big|_d = -k_d\,[I] = -r_d. \tag{2.72}$$

The rate coefficient is calculated by the Arrhenius equation. Kinetical data typically contains the activation energy $E_{a,d}$ and - instead of the pre-exponential factor - the temperature T_{10h}, at which the initiator has a half-life of ten hours

$$A_d = \frac{\ln 2}{36000s}\,e^{\frac{E_{a,d}}{\Re T_{10h}}}. \tag{2.73}$$

Whereas two primary radicals are formed from an initiator molecule, chain initiation is only one kind of different possible reactions of theses radicals (Hutchinson 2005, p. 155; Lechner, Gehrke, and Nordmeier 1996, p. 54). The rather complicated dependency of these processes on the reaction system is commonly simplified by a fractional initiator efficiency f_d, with $0 \le f_d \le 1$. A living chain of length one is created by reaction of an initiator radical with a monomer molecule. This step is typically so fast, that initiator radicals are, independently from the monomer concentration, almost instantly consumed. In comparison, initiator decomposition is much slower and therefore the time-determining step considering both consecutive reactions. Hence, a quasi-steady state approximation can be adopted for the initiator radicals, as long as monomer is present in the system in abundance. Then monomer consumption by chain initiation is directly linked to the initiator concentration without the need of balancing the initiator radical

$$r_{I^*}^F\big|_d = 2f_d r_d - k_p\,[I^*]\,[M] \approx 0 \tag{2.74}$$
$$r_M^F\big|_i = -k_p\,[I^*]\,[M] \approx -2f_d r_d \tag{2.75}$$
$$r_{R_1}^F\big|_i = k_p\,[I^*]\,[M] \approx 2f_d r_d. \tag{2.76}$$

Accordingly, the first two reactions in Table 2.2 are commonly condensed into one single step.

Chain Propagation

Chain propagation is the main reaction in free radical polymerisation. A monomer molecule attaches at the radical site of the chain. The radical is transferred to this monomer unit and the chain grows. If polymer of high molecular weight is being produced, monomer will be virtually exclusively consumed by chain propagation (Hutchinson 2005, p. 159). Commonly, this long-chain hypothesis (LCH) is valid. Monomer consumption by initiation or transfer reactions is practically negligible. Furthermore, the total heat of reaction is virtually solely generated by chain propagation. The amount of living chains of length s shrinks due to propagation reactions involving R_s molecules and grows by propagation of R_{s-1} chains. Rates of formation for living chains and monomer are

$$r_{R_s}^F\big|_p = k_p \left([R_{s-1}] - [R_s]\right) \tag{2.77}$$

$$r_M^F\big|_p = -k_p \sum_s [R_s] = -k_p [R_{tot}], \tag{2.78}$$

with $[R_{tot}]$ being the total concentration of living chains.

Termination

If two living chains react with each other, their radicals will be deactivated and chain propagation stopped. Both chains either remain seperately as two dead chains (termination by disproportionation) or combine to one single dead chain (termination by recombination). With respect to the consumption of living chains, both possibilities are mathematically identical, whereas the rate of formation of dead chains depends on the kind of termination

$$r_{R_s}^F\big|_{td+tc} = (k_{td} + k_{tc}) [R_s] \sum_t [R_t] = (k_{td} + k_{tc}) [R_s] [R_{tot}] \tag{2.79}$$

$$r_{P_s}^F\big|_{td} = k_{td} [R_s] \sum_t [R_t] = k_{td} [R_s] [R_{tot}] \tag{2.80}$$

$$r_{P_s}^F\big|_{tc} = k_{tc} \sum_{t=1}^{s-1} [R_t] [R_{s-t}]. \tag{2.81}$$

Transfer to Monomer or Solvent/Transfer Agent

Along propagation, a living chain may react with a monomer molecule and transfer its radical to the monomer without addition of this monomer unit to the chain.

In doing so, the living chain is terminated and the monomer becomes a living chain of length one. Whereas this reaction is by far less probable than propagation, it affects the chain length distributions of living and dead chains. The resulting living chain only contains one monomer unit, thus reducing the average chain length of the living polymer species and thus the degree of polymerisation of dead chains. Similarly, living chains can react with solvent or a special transfer agent, which can be used in order to control/limit the polymer's molecular weight. In most cases, solvent or transfer agent radicals are short-lived, and it can be assumed that the initiation of a new chain by these radicals has a similar time scale as one propagation step (Hutchinson 2005, p. 167)

$$r_{R_s}^F\big|_{tr} = -k_{trm}[M][R_s] - k_{trs}[S][R_s] \tag{2.82}$$

$$r_{R_1}^F\big|_{tr} = k_{trm}[M]\sum_s[R_s] + k_{trs}[S]\sum_s[R_s] \tag{2.83}$$

$$r_{P_s}^F\big|_{tr} = k_{trm}[M][R_s] + k_{trs}[S][R_s] \tag{2.84}$$

$$r_M^F\big|_{tr} = -k_{trm}[M]\sum_s[R_s] - k_{trs}[S]\sum_s[R_s]. \tag{2.85}$$

Transfer to Polymer

If a living chain reacts with an already terminated polymer chain, it can transfer its radical to one of the monomer units of the dead chain. The radical is typically not situated at the end of the reactivated chain and further propagation reactions result in the formation of a side-chain. Accordingly, this transfer mechanism is often reffered to as long-chain branching. Furthermore, as the number of possible monomer units which may be attacked is proportional to the length of the dead chain, the probability of this reaction is increased with a higher degree of polymerisation. Unlike in most other polymerisation reactions, the reaction rate is therefore depending on the chain length of the reacting dead chain

$$r_{R_s}^F\big|_{trp} = -k_{trp}[R_s]\sum_t t[P_t] + k_{trp}s[P_s]\sum_t[R_t] \tag{2.86}$$

$$r_{P_s}^F\big|_{trp} = k_{trp}[R_s]\sum_t t[P_t] - k_{trp}s[P_s]\sum_t[R_t]. \tag{2.87}$$

The notation of branched polymers is simplified here. Often an additional branching index is added to the polymer species, indicating for the number of side-chains in the polymer molecule. A living or dead polymer $R_{s,t}$ and $P_{s,t}$ then denotes a polymer molecule of length s with t branches.

2.3.2 Quasi-Steady-State Assumption (QSSA)

In free radical polymerisation, there is almost instantly an equilibrium between the generation of new radicals and the termination of chains. Due to the fast dynamics of the radical reactions in comparison to the polymerisation system at all, a quasi-steady-state assumption for the total concentration of living chains can be applied (Hutchinson 2005, p. 159)

$$r_i = r_{td} + r_{tc} \qquad (2.88)$$

$$[R_{tot}] = \sqrt{\frac{2fk_d[I]}{(k_{td}+k_{tc})}}. \qquad (2.89)$$

Furthermore, the kinetic chain-length v (the average length of living chains) can be obtained from the ratio of consumed monomer due to propagation to the number of terminated or due to transfer reactions reactivated living chains. The current degree of polymerisation DP^{inst} of the chains generated at an instant is calculated in an analogous way. However, the number of originating dead chains by termination is dependent on the kind of termination (Hutchinson 2005, p. 159)

$$v = \frac{r_p}{r_t + r_{tr}} = \frac{k_p[M]}{(k_{td}+k_{tc})[R_{tot}]+k_{trm}[M]+k_{trs}[S]} \qquad (2.90)$$

$$DP^{inst} = \frac{r_p}{r_{td}+0.5r_{tc}+r_{tr}} = \frac{k_p[M]}{(k_{td}+0.5k_{tc})[R_{tot}]+k_{trm}[M]+k_{trs}[S]}. \qquad (2.91)$$

Similar expressions can be derived for copolymerisation systems. Following the QSSA, the radical concentration as well as the instant number averages of the chain length distributions of living and dead chains can be obtained solely from the educt concentrations, namely initiator and monomer, by means of algebraic equations. Through this, it is a very efficient approach if only number averaged values need to be obtained and the primary polymerisation reactions are propagation, termination and transfer to monomer/solvent. The (cumulated) average molar mass of the polymer can be evaluated, if both the concentration and the partial density of dead chains are balanced as components, as well. More complicated reactions, such as transfer to polymer, cannot be considered. Besides, this approach does not provide any information on the weight averages of the chain length distribution and, accordingly, on the dispersity of the generated polymer. It is, however, possible to obtain these values, under the presumption that there are only linear chains (Hutchinson 2005, p. 201).

2.3.3 Method of Moments

Generally, statistical moments λ_k^f of a distribution of an entity $f(s)$, depending on a variable s, are defined as

$$\lambda_k^f = \int_{s_{min}}^{s_{max}} s^k f(s)\, ds, \tag{2.92}$$

where k is the index of the moment. Unlike other systems, such as droplet size distributions in a spray, the chain length distributions of polymers are not based upon a continuous variable, but on the discrete number of monomer units s in the chains. Thus, the integral is changed to a summation over all possible chain-length with step-size one. The k-th moments of the chain-length distributions of dead (P, ζ_k) and living (R, λ_k) chains are then expressed as follows

$$\zeta_k = \sum_{s=1}^{\infty} s^k [P_s] \tag{2.93}$$

$$\lambda_k = \sum_{s=1}^{\infty} s^k [R_s]. \tag{2.94}$$

The zeroth moment is just the summation of the concentrations of all chains and therefore denotes the total concentration of polymer, whereas the first moment yields the number of monomer units in all chains. Higher moments do not have a distinct meaning with respect to the polymeric system. Number (P_n), mass (P_m) and Z averages (P_z) of a chain length distribution are obtained by

$$P_n = \frac{\text{number of monomer units in chains}}{\text{number of chains}} = \frac{\sum_s s\,[P_s]}{\sum_s [P_s]} = \frac{\zeta_1}{\zeta_0} \tag{2.95}$$

$$P_w = \frac{\sum_s s^2\,[P_s]}{\sum_s s\,[P_s]} = \frac{\zeta_2}{\zeta_1} \tag{2.96}$$

$$P_z = \frac{\sum_s s^3\,[P_s]}{\sum_s s^2\,[P_s]} = \frac{\zeta_3}{\zeta_2}. \tag{2.97}$$

The dispersity $Đ_X$ (formerly polydispersity index, PDI) is

$$Đ_X = \frac{P_w}{P_n} = \frac{\zeta_0 \zeta_2}{\zeta_1^2}. \tag{2.98}$$

The moments' reaction rates are obtained from the rate of formation considering a chain length s, subsequent multiplication with s^k and summation over all s.

As an example, the moments' formation rates with respect to termination by combination can be derived from the differential equations of living chains of size s as follows:

$$\left. \frac{d[R_s]}{dt} \right|_{tc} = -k_{tc}[R_s]\sum_t [R_t] \tag{2.99}$$

$$\sum_{s=1}^{\infty} s^k \left. \frac{d[R_s]}{dt} \right|_{tc} = -k_{tc}\sum_{s=1}^{\infty} s^k[R_s]\sum_t t^0[R_t] \tag{2.100}$$

$$\left. r_{\lambda_k}^F \right|_{tc} = \left. \frac{d\lambda_k}{dt} \right|_{tc} = -k_{tc}\lambda_0\lambda_k. \tag{2.101}$$

If the k-th moment only depends on moments of order k or lower, moment equations are closed. If, however, the number of unknown moments exceeds the number of equations, the system is not closed and additional assumptions have to be met. An example is transfer to polymer, in which the reaction rate depends on the chain length of the dead chain. In

$$\sum s^k \left. \frac{d[R_s]}{dt} \right|_{trp} = -k_{trp}\sum s^k[R_s]\sum_t t[P_t] + k_{trp}\sum s^k s[P_s]\sum_t [R_t]$$

$$\left. \frac{d\lambda_k}{dt} \right|_{trp} = k_{trp}(-\lambda_k\zeta_1 + \zeta_{k+1}\lambda_0) \tag{2.102}$$

$$\left. \frac{d\zeta_k}{dt} \right|_{trp} = k_{trp}(\lambda_k\zeta_1 - \zeta_{k+1}\lambda_0) \tag{2.103}$$

the $k+1$-th moment of the distribution of dead chains is required for the calculation of the k-th moment of living and dead chains' distributions. Commonly, the moments of order zero to two are balanced in a mathematical model and the third moment ζ_3 needs to be obtained by a closing condition. Hulburt and Katz (1964) expressed the moments of a distribution by a Laguerre polynomial. Under the condition that this polynomial can be truncated after the first term with sufficient accuracy, this term provides a closing condition based on the first three moments ζ_0, ζ_1 und ζ_2, which can be applied to all remaining moments of the distribution. The third moment is then

$$\zeta_3 = \frac{\zeta_2}{\zeta_0\zeta_1}\left(2\zeta_0\zeta_2 - \zeta_1^2\right). \tag{2.104}$$

The moment formation rates with respect to the distributions of living and dead chains are summarised in Table 2.3.

Table 2.3: Formation rates of statistical moments in free radical homopolymerisation for initiation, propagation, termination by disproportionation or combination and transfer to monomer and polymer.

moment	rate of formation
λ_0	$2fk_d[I] - (k_{td} + k_{tc})\lambda_0^2$
λ_1	$2fk_d[I] + k_p[M]\lambda_0 - (k_{td} + k_{tc})\lambda_0\lambda_1 - k_{trm}[M](\lambda_1 - \lambda_0)$
	$-k_{trp}(\lambda_1\zeta_1 - \lambda_0\zeta_2)$
λ_2	$2fk_d[I] + k_p[M](2\lambda_1 + \lambda_0) - (k_{td} + k_{tc})\lambda_0\lambda_2 - k_{trm}[M](\lambda_2 - \lambda_0)$
	$-k_{trp}(\lambda_2\zeta_1 - \lambda_0\zeta_3)$
ζ_0	$(k_{td} + 0.5k_{tc})\lambda_0^2 + k_{trm}[M]\lambda_0$
ζ_1	$(k_{td} + k_{tc})\lambda_0\lambda_1 + k_{trm}[M]\lambda_1 + k_{trp}(\lambda_1\zeta_1 - \lambda_0\zeta_2)$
ζ_2	$(k_{td} + k_{tc})\lambda_0\lambda_2 + k_{tc}\lambda_1^2 + k_{trm}[M]\lambda_2 + k_{trp}(\lambda_2\zeta_1 - \lambda_0\zeta_3)$

A different approach in order to close the moments are bulk moments, which have been introduced by Arriola (1989, p. 12) and refer to the combined distribution of both living and dead chains

$$\mu_k = \lambda_k + \zeta_k \approx \zeta_k. \tag{2.105}$$

Reaction rates of bulk moments (Table 2.4) can simply be obtained by addition of corresponding expressions for λ_k and ζ_k. The terms of transfer to polymer cancel each other out. After a negligible starting-up phase $\zeta_k \gg \lambda_k$. The moments of the dead chains are virtually identical to the bulk moments and can be replaced by those when calculating the reaction rates of living chains' moments. Furthermore the calculation of bulk moments only requires the moments of the living chains up to first order. Hence, the moment equations are closed with five equations providing full information about the statistical values of the obtained polymer (Hutchinson 2005, p. 201). If the truncation of the Laguerre polynomial does not significantly affect the accuracy of approximation of the moment's distribution, both approaches for closure will provide comparable results. This has, for instance, been shown by Baltsas, Achilias, and Kiparissides (1996) for the application of a copolymerisation with long-chain-branching in a CSTR.

The method of moments offers several advantages. The complicated system of polymerisation reactions can be expressed by only six, in case of bulk moments five, differential equations for the moments. Thereby the method is

Table 2.4: Reaction rates of bulk moments in free radical homopolymerisation.

moment	rate of formation
μ_0	$2fk_d\,[I] - 0.5k_{tc}\lambda_0{}^2 + k_{trm}\,[M]\,\lambda_0$
μ_1	$2fk_d\,[I] + k_p\,[M]\,\lambda_0 + k_{trm}\,[M]\,\lambda_1$
μ_2	$2fk_d\,[I] + k_p\,[M]\,(2\lambda_1 + \lambda_0) + k_{tc}\lambda_1{}^2 + k_{trm}\,[M]\,\lambda_0$

very fast and comparably easy to implement. In comparison to the QSSA not only the number average of a chain length distributions, but the mass average and polydispersity can be obtained as well. Furthermore, it is very versatile and a great variety of different polymerisation reaction mechanisms can be implemented. Additional properties like the distribution of incorporated monomer of different kind in copolymerisation or the number of branches or crosslinks can be considered using additional moment indices. The computed values are mathematically exact, as long as the moment equations are closed. The complete curve of the chain length distribution cannot be obtained from the statistical moments. Number and mass average values are often of main interest, though, and provided by the method of moments in a consistent manner.

2.4 Mixture Thermodynamics

Thermodynamics affect the evaporation of volatile components and Maxwell-Stefan diffusion. Mixed components may behave strongly different to pure species. In ideal mixtures, physical values or effects are just "scaled" by the mole fractions of the single species, e.g. one component's vapour pressure at the droplet's surface behaves proportional to its mole fraction. This corresponds with the lower surface coverage of this species in a mixture and is sensible if molar volumes of different components are similar. Macromolecules, by contrast, occupy a much larger part of the droplet surface area than their fraction of the total mole number suggests. In case of spray polymerisation, the assumption of ideality is therefore not sufficient. Mixture thermodynamics need to be taken into account and directly affect mass transfer at the droplet surface. Concerning diffusion, thermodynamics are required in Maxwell-Stefan equations. If generalised or pseudo-binary Fick's laws are applied, the concentration dependency of the diffusion coefficients implicitly includes thermodynamical effects.

2.4.1 Vapour Liquid Equilibrium at the Droplet's Surface

At the interface between liquid (superscript L) and gas (G), there is thermodynamic equilibrium between both phases

$$\mu_j^L = \mu_j^G \tag{2.106}$$

$$\gamma_j x_j f_j^0 = \varphi_j p_j^G. \tag{2.107}$$

γ_j is the activity coefficient of compontent j with $a_j = \gamma_j x_j$ defining the thermodynamic activity a_j. f_j^0 and φ_j are the liquid phase reference fugacity and the vapour phase fugacity coefficient, respectively. p_j^G denotes the partial pressure of species j in the gas phase, directly at the liquid surface. The vapour phase fugacity coefficient accounts for the non-ideality of the gas phase and is typically derived by a virial equation. The reference fugacity f_j^0 of an incompressible liquid (no pressure dependence of the molar volume v_j) is defined as

$$f_j^0 = \varphi^S p^S (T) e^{\frac{v_j \left(p - p^S \right)}{\Re T}} \tag{2.108}$$

p^S is the saturation pressure. The exponential term is also known as the Poynting factor. At moderate pressures, it typically does not vary significantly from one and can be neglected. φ^S is the vapour phase fugacity of component j at saturation, again accounting for the non-ideality of the gas-phase. Assuming ideal gas and neglecting the Poynting factor, the vapour liquid equilibrium at the droplet surface

$$p_j^G = \gamma_j x_j p^S (T) \tag{2.109}$$

provides an equation for the partial pressure of species j, which can be further applied in a linear driving force approach for mass transfer. The only value accounting for non-ideality is the activity coefficient γ_j. In case of ideal mixture thermodynamics in the liquid, $\gamma_j = 1$ and equation 2.109 further simplifies to Raoult's law

$$p_j^G = x_j p^S (T). \tag{2.110}$$

The saturation pressure can be derived by correlations like the ones of Clausius-Clapeyron or the different kinds of Antoine's equation. In this work, the three parameter Antoine equation has been applied:

$$\log_{10} p^S = A - \frac{B}{T+C}. \tag{2.111}$$

2.4.2 Calculation of Activity Coefficients

The activity coefficient γ_j accounts for the non-ideal behaviour of component j in a liquid mixture, whereas ideality corrsponds with $\gamma_j = 1$. The literature provides a variety of different thermodynamic models for the calculation of activities in a mixture based on theoretical approaches and experimentally obtained parameters. The Flory-Huggins model was one of the first theories to account for the non-ideality of polymer mixtures (Flory 1942; Huggins 1941). It predicts activity coefficients according to the molar volume of molecules and an additional parameter accounting for interaction between different kinds of molecules. In the 1970s, group contribution methods became popular, in which the single molecules in a mixture are considered as aggregates of different functional groups. The basic presumption of these methods is that the overall thermodynamic behaviour in a mixture can be derived from characteristics of these basic groups and interactions amongst them. Whereas group properties, such as van der Waals volumes and surfaces, are typically kept fixed, interaction parameters have to be derived by experiments and their values are frequently updated. The charm of such models lies in their fairly good ability of extrapolating binary data. Whereas experimental data is often available for binary mixtures, it is elaborate to obtain reliable values for mixtures containing three or more components. Group contribution models tuned to binary experiments have proven to provide reasonably good results, when adapting the corresponding group interaction parameters to polynary mixtures. Popular group contribution models are UNIQUAC (UNIversal QUAsi Chemical, Abrams and Prausnitz 1975) and especially UNIFAC (UNIversal quasichemical Functional group Activity Coefficients, Fredenslund, Jones, and Prausnitz 1975) and its various descendents. In the context of polymer thermodynamics the PC-SAFT equation of state (Perturbed-Chain Statistical Associating Fluid Theory, Gross and Sadowski 2001) provides very good data, but is very complex and may rather be applied by external software. An ordinary UNIFAC implementation is applied within this work for reasons of simplicity.

2.4.3 The UNIFAC Equations

UNIFAC and its variants belong to the class of so called g^E models, which relate the activity coefficients to the excess Gibbs free energy (Abrams and Prausnitz 1975)

$$ng^E = \Re T \sum_j n_j \ln \gamma_j \qquad \longrightarrow \qquad \Re T \ln \gamma_j = \frac{\partial \left(ng^E \right)}{\partial n_j} \bigg|_{T,p,n_{i \neq j}} . \qquad (2.112)$$

The activity coefficient γ_j of a species j in a mixture is described by

$$\ln \gamma_j = \ln \gamma_j^C + \ln \gamma_j^R \qquad (2.113)$$

The combinatorial part γ_j^C accounts for size and shape effects. It only depends on the van-der-Waals volumes and surfaces of the single groups and is temperature independent. Energetic group interactions are considered by the residual contribution γ_j^R, which is determined by experimentally adjusted group interaction parameters and the temperature.

The combinatorial contribution corresponds with the older UNIQUAC model (Abrams and Prausnitz 1975; Fredenslund, Gmehling, and Rasmussen 1977, pp. 24, 31)

$$\ln \gamma_j^C = \ln \frac{\Phi_j}{x_j} + \frac{z}{2} q_j \ln \frac{\theta_j}{\Phi_j} + l_j - \frac{\Phi_j}{x_j} \sum_k x_k l_k. \qquad (2.114)$$

$$l_j = \frac{z}{2} (r_j - q_j) - (r_j - 1) \qquad \theta_j = \frac{q_j x_j}{\sum_k q_k x_k} \qquad \Phi_j = \frac{r_j x_j}{\sum_k r_k x_k}.$$

The coordination number z is generally chosen to 10. Φ and θ are the molecular volume and surface area fractions, respectively. The molecular volumes and surface areas r_j and q_j are defined as

$$r_j = \sum_k v_k^{(j)} R_k \qquad q_j = \sum_k v_k^{(j)} Q_k \qquad (2.115)$$

based on the volumes R_j and surface areas Q_j of the single groups within a molecule, with $v_k^{(j)}$ being the number of groups k in a molecule j. These group values rely on measured van der Waals group volumes and surface areas, taken from Bondi (1968), which are normalised with respect to the standard segment values of a single CH_2 group in a polyethylene molecule of infinite length (Abrams and Prausnitz 1975). The residual contribution

$$\ln \gamma_j^R = \sum_k v_k^{(j)} \left(\ln \Gamma_k - \ln \Gamma_k^{(j)} \right) \qquad (2.116)$$

relies on the residual activity coefficients Γ_k of group k in a solution (Fredenslund, Gmehling, and Rasmussen 1977, p. 28), with $\Gamma_k^{(j)}$ being the corresponding activity coefficient in pure component j

$$\ln \Gamma_k = Q_k \left[1 - \ln\left(\sum_j \Theta_m \Psi_{mk} \right) - \sum_m \left(\frac{\Theta_m \Psi_{km}}{\sum_n \Theta_n \Psi_{nm}} \right) \right] \qquad (2.117)$$

$$\Theta_m = \frac{Q_m X_m}{\sum_n Q_n X_n} \qquad X_m = \frac{\sum_j v_m^{(j)} x_j}{\sum_j \sum_n v_n^{(j)} x_j} \qquad \Psi_{nm} = e^{-\frac{a_{nm}}{T}}.$$

Θ_m and X_m denote the group surface area fraction and the group fraction in the mixture, respectively, whereas the parameter Ψ_{nm} accounts for the temperature dependent interaction between groups, based on the group-interaction parameter a_{nm}, which essentially is an activation energy (with the division by $\mathfrak{R}$ already incorporated into the value so that it has the unit of a temperature).

In the UNIFAC model, sub-groups and main groups are distinguished. Sub-groups within a main group are of equivalent kind so that they do not show considerable energetic interactions amongst themselves. Hence, only different main groups contribute to the residual part γ_j^R. However, the sub-groups are of different volume and surface area - like, as an example, the chain segments C, CH, CH_2 and CH_3 of the first main group CH_2. Hence the combinatorial contribution depends on the distinct sub-group volumes and surface areas.

2.5 Spray Drying: Basic Assumptions and Physical Effects with Respect to Single Droplets

Mathematical modelling in this work is focused on processes within droplets. Single droplet drying models generally apply a number of simplifying assumptions. Heat and mass transfer are typically based on linear driving force approaches based on dimensionless numbers rather than on resolving the respective transport equations in the gas in detail. The effect of inner circulation inside the droplet is typically neglected. In the following these common approaches in single droplet modelling and some basic estimations of the whole drying process will be introduced.

2.5.1 Approximate Residence Time in a Spray Dryer

The three-dimensional flow field in a spray dryer is complicated and depends on various parameters like incoming gas fluxes, droplet atomisation, operation mode (co- or counter-current) and the geometry of the spray dryer. The residence time of a droplet within a dryer therefore not only depends on its size but may also differ locally and over time. As a very rough rule of thumb, the residence time of a droplet with radius R in a spray dryer of height H can be estimated by Stokes' relation for the terminal velocity of a spherical object within stagnant air (Stokes 1851):

$$v(R) = \frac{2g}{9\eta^G} \left(\rho^L - \rho^G \right) R^2 \qquad (2.118)$$

$$t(R) = \frac{H}{v} = \frac{9\eta^G}{2g\left(\rho^L - \rho^G\right)} \frac{H}{R^2}. \qquad (2.119)$$

g is the force of gravity and the superscripts L and G denote the liquid and gas phase, respectively. Gunn and Kinzer (1949) found that the terminal velocity of water droplets in stagnant air corresponds with Stokes' solution for droplet radii smaller than approximately $50\,\mu m$. In case of larger droplets it is overestimated by Stokes' law. Figure 2.2 provides the correlation of falling time, height and droplet radius obtained by Stokes' equation for the terminal velocity

Residence time distributions have been measured by Kieviet and Kerkhof (1995) and Mazza, Brandão, and Wildhagen (2003). A particle size distribution with $d_{w,50} = 134\,\mu m$ exhibited a median residence time of 58.5 s within a pilot-plant co-current spray dryer of $\approx 3.7\,m$ height in the experiments of Kieviet and Kerkhof. Mazza, Brandão, and Wildhagen measured an estimated average residence time of 72.5 s for final particles of about $4.4\,\mu m$ volumetric median diameter in a pilot-plant dryer of about 1.8 m height. Both investigations found that the residence times of droplets may differ significantly to the one of the drying gas.

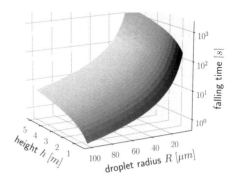

Figure 2.2: Falling time according to Stokes' law in dependence of height and droplet radius.

Reactive drying processes do not only necessitate completion of product drying but also of chemical reactions. In comparison to batch, continuous stirred tank or tube reactors, the residence time of a drop therefore sets a limitation to processes, which are often carried out on considerably larger time scales. With respect to spray polymerisation, this process route is only feasable for fast polymerisation reactions such as poly acrylic acid synthesis. When discussing simulation results, residence times can be presumed in the order of magnitude of less than ten to one hundred seconds, depending on the droplet size.

2.5.2 Heat and Mass Transfer

As in many other applications, heat and mass transfer in spray drying are typically approximated by averaged approximate rules based on the dimensionless Nusselt and Sherwood numbers. The contributions of convection and diffusion within a boundary layer surrounding the droplet are boiled down to heat or mass transfer coefficients acting on linear driving forces. In doing so, the drying behaviour can be well represented and the models are greatly simplified, without the need of detailed knowledge on a local flowfield. The heat flux based on a heat transfer coefficient α is defined as (Baehr and Stephan 2010, p. 12)

$$\vec{q}^{\Gamma} = \alpha \left(T^{surf} - T^{\infty} \right) \vec{n}, \tag{2.120}$$

where T^{∞} is he temperature outside the boundary layer (in this case around the droplet) and T^{surf} the surface temperature. $\vec{n}$ is the unit normal towards the interface, pointing from the surface into the boundary layer. The superscript Γ denotes that an interfacial flux is calculated. In the same way the mass flux over an interface is expressed as (Baehr and Stephan 2010, p. 86)

$$\vec{\Omega}_j^N = \vec{J}_j^{\Gamma,N} = \beta \left(c_j^{surf} - c_j^{\infty} \right) \vec{n} \overset{ideal\ gas}{=} \frac{\beta}{\Re \bar{T}} \left(p_j^{surf} - p_j^{\infty} \right) \vec{n}, \tag{2.121}$$

in which β is the mass transfer coefficient and the superscripts ^{surf} and $^{\infty}$ again denote the conditions on the surface and outside the boundary layer, respectively. In the following, mass and molar fluxes across the interface between droplet and drying gas will be written as $\vec{\Omega}_j$ and $\vec{\Omega}_j^N$, respectively. When expressed in terms of partial pressures on the right hand side of equation 2.121, the temperature $\bar{T}$ is averaged over the boundary layer. The heat and mass transfer coefficients are expressed by the corresponding heat and mass conductivities λ^G and D^G and the extend of the boundary layer. This geometric factor L can be considered as the length scale of the problem. Additionally, the effect of the flowfield withing the geometry is taken into account by the dimensionless Nusselt Nu and Sherwood Sh numbers (Baehr and Stephan 2010, pp. 20, 88)

$$\alpha = \frac{\lambda^G}{L} Nu \qquad (2.122)$$

$$\beta = \frac{D^G}{L} Sh. \qquad (2.123)$$

In case of a droplet, L equals the droplet diameter. There are many (slightly) different Nusselt and Sherwood correlations in the literature of droplet drying, Typically, these depend on the Reynolds and Prandtl or Schmidt numbers, like the ones being proposed by Ranz and Marshall (1952a,b):

$$Nu = 2 + 0.6 Pr^{1/3} Re^{1/2} \qquad (2.124)$$

$$Sh = 2 + 0.6 Sc^{1/3} Re^{1/2}. \qquad (2.125)$$

A value of 2 is the lower limit of both numbers and corresponds with heat and mass transfer in stagnant air. For the main purpose of this work, modelling of spray polymerisation and the development of a model considering structure evolution, the differences in calculating the Nusselt and Sherwood numbers in various models are of minor interest. For reasons of simplicity, Nusselt and Sherwood numbers are presupposed depending on the initial droplet geometry and kept fixed throughout the computations in this work. The fluid within the boundary layer is assumed as an ideal gas as on the right hand side of equation 2.121. The averaged temperature inside the boundary layer $\bar{T}$ is obtained using the surface temperature and T^∞ via an arithemtic mean.

2.5.3   Inner Circulation Inside a Droplet

The assumption of ideal sphericity neglects transport in azimuthal direction. However, due to friction at the droplet surface the gas flow around the drop will induce a circular liquid flux inside the droplet. The impact of this effect has been estimated by Muginstein, Fichman, and Gutfinger (2001). The liquid velocity tangential to the droplet surface v^L corresponds to the relative velocity of the droplet to the gas and its radius R by

$$v^L = \frac{\eta^G}{\eta^G + \eta^L} (1 - 2R) v^G \approx \frac{1}{50} v^G. \qquad (2.126)$$

If diffusive mass transfer is intense, convective mixing inside the drop is subordinate. Muginstein, Fichman, and Gutfinger found that the effect of inner circulation can be neglected if the Péclet number of the liquid near the droplet surface

$$Pe^L = \frac{v^L \delta^L}{D^L} \tag{2.127}$$

is smaller than 100, with D^L being the diffusion coefficient in the liquid and δ^L the thickness of the thin liquid film at the droplet surface. Stokes' equation 2.118 overestimates the velocity relative to the gas for large droplets and is therefore conservative with respect to lateral liquid motion inside the drop. Assuming this film being at least one order of magnitude smaller than the droplet radius, the minimum diffusion coefficient, at which the effect of inner circulation is negligible, is

$$D^L > \frac{v^L 0.1 R}{100} \approx \frac{g \left(\rho^L - \rho^G\right) R^3}{225000 \eta^G}. \tag{2.128}$$

Droplet diameters of 50, 75 and 100 µm yield minimum diffusion coefficients of about 0.26, 0.86 and 2.08×10^{-9} m^2/s. With diffusion coefficients in liquid mixtures being in the order of 10^{-9} m^2/s, an intensification of mass transfer due to internal mixing might occur for droplet radii larger then 75 µm. On the other hand, Stokes' equation overestimates the terminal velocity for droplets of this size that the effect will still be comparably small. Mixtures containing a significant amount of polymer exhibit much smaller diffusion coefficient, caused by an increase of fluid viscosity. In the first approximation, both sides of the equation will change similarly so that inner mixing is not affected and the Peclet number remains in the same order of magnitude as before.

Concerning single droplet experiments in an ultrasound levitator, liquid motion may be intensified. According to Brenn et al. (2007), acoustic streaming around the droplet induces additional convection inside the droplet and a fully mixed regime is to be assumed. On the contrary, Sloth et al. (2006) employ the approach of Muginstein, Fichman, and Gutfinger with the result that solely radial gradients have to be accounted for. Both contributions provide model results in accordance with experimental findings, which makes it difficult to provide a general statement. Application of the models in this work to ultrasound levitator experiments may therefore necessitate further consideration of additional convection inside the droplet. For droplet motion within a gas, especially the case of small droplets, inner circulation can be considered as negligible.

2.5.4 Are Droplets Fully Mixed?

Even if homogenisation due to inner circulation needs not to be taken into account, diffusive transport may lead to complete mixing of the droplet with respect to either heat or mass transport - or both. The Biot numbers (subscript m for mass transfer) set the diffusivities on both sides of an interface into relation and provide therefore a measure for diffusive equilibration within the droplet for given heat and mass transfer coefficients (Baehr and Stephan 2010, p. 130)

$$Bi = \frac{\alpha R}{\lambda^L} = \frac{\lambda^G Nu}{\lambda^L \cdot 2} \tag{2.129}$$

$$Bi_m = \frac{\beta R}{D^L} = \frac{D^G Sh}{D^L \cdot 2}. \tag{2.130}$$

A heat transport Biot number smaller than 0.1 denotes that diffusive transport within an object is fast compared to outer transport so that a lumped approximation of the whole body without resolving inner transport is feasible (Baehr and Stephan 2010, p. 196). Insertion of approximate values for nitrogen and water (the order of magnitude is relevant)

$$Bi \approx \frac{0.03}{0.6} \frac{Nu}{2} = 0.025 Nu \tag{2.131}$$

$$Bi_m \approx \frac{10^{-5}}{10^{-9}} \frac{Sh}{2} = 5000 Sh, \tag{2.132}$$

the droplet can be assumed as fully mixed with respect to heat transport as long as the Nusselt number is fairly low. Parti (1994) points out that despite the otherwise valid assumption of equality of heat and mass transfer, the Biot mass number cannot be interpreted similar to the heat transfer Biot number. As concentrations and their gradients differ between liquid and gas by orders of magnitude, the mass flux over an interface can be small even if the mass transfer coefficient is comparably high. Consequently, a problem may still be treated in a lumped way even if the mass transfer Biot number exceeds the limit value of 0.1. Considering maximum water concentrations in liquid of about $55.5 \times 10^3 \, mol/m^3$ and $33 \, mol/m^3$ in the drying gas (at $100 \, °C$) the ratio between liquid and gas concentrations is about 1700. Hence, the limit mass transfer Biot number is in the order of 170 and values below indicate full mixing of the droplet. Hence, the approximated Bi_m indicates that mass transport needs to be modelled in detail and that lumped modelling will introduce errors. Yet, the effect is not as large as the very high mass transfer Biot number suggests at first glance.

3. Single Droplet Modelling of Solution Drying, Reactive Drying and Spray Polymerisation

Spray drying is a very common process in industries. Its wide span of applications is reflected in a variety of different kinds of mathematical models. Free radical polymerisation is typically carried out in a solution. Model equations are therefore limited to a solution of one single, quasi-homogeneous phase in this work. A disperse phase is not considered. Spherically symmetric droplets are presumed, which is common practice for spray drying in the literature and feasible regarding the considerations in sections 2.5.3 and 2.5.4. Spatial gradients are hence limited to the radial direction. It is possible to enhance the methodology presented here to cases, in which a solid crust occurs, by implementation of an additional continuous phase and a further interface between both phases.

Typical literature models for single droplet drying employ transport equations for mass fractions. In this work, model equations are derived in both mass and molar notation for several reasons. In case of drying without reactions, it is appropriate to use mass fractions or partial densities, as one is rather interested in the mass, which has evaporated from a droplet, than in the number of moles or molecules. Moreover, the feed material is more easily described by weighted portions. Easy chemical reactions can be modelled using a mass based description as well. In case of a non-ideal mixture behaviour or complicated reactions like spray polymerisation, however, a molar formulation is favourable. The statistical moments of the chain length distribution are concentrations, which makes it reasonable to calculate all species using a molar description. Furthermore, the diffusive behaviour of the polymeric system is not trivial, even for the simple approach of (pseudo-)binary Fickian diffusion, as will be shown in sections 3.1.4 and 3.1.5. When modelling higher moments or the molar weight, polymer diffusion needs to be considered with respect to the single chain molecules rather than a polymer partial density.

First of all, a lumped model of (reactive) solution drying processes will be discussed shortly, which does not account for transport limitations inside the droplet and assumes an ideal drying behaviour. After this, a one-dimensional model will be derived, considering drying of one quasi-homogeneous phase with additional chemical reactions. This model is valid either for drying of a solution without reactions or reactive drying processes with constant physical properties of the concerned species. Except for special cases like spray polymerisation, it covers the majority of imaginable reactive drying processes. Subsequently, a further developed reactive drying model will be discussed, which accounts for the peculiarities of spray polymerisation, namely the non-constant averaged molar mass of the polymer. At last, the implementation of single droplet reactive drying models into a numerical code will be discussed.

3.1 Transport in a Reaction-Diffusion System - Diffusion and Reaction Driven Convection

In a mixture of various species having different densities, the average density varies according to changes in the mixture composition over space or time. Diffusion of species with different densities then leads to a change in the overall density and, considering the continuity equation 2.3 or 2.28, to a non-zero velocity field. Considering two control volumes, the diffusive volume fluxes between these do not necessarily add up to zero (as the corresponding molar or mass fluxes do) - depending on the specific volumes ($\frac{1}{\rho_j}$ or $\frac{MW_j}{\rho_j}$) of the respective species. As a consequence one of these volumes would expand, whereas the other element would shrink. The diffusion induced convective flux corrects this in such a way, that both elements preserve their volume and the continuity equation is fulfilled. A similar effect occurs in case of volume or density changing chemical reactions. If the density of a reaction product within a droplet is higher than the educt's value, the droplet will shrink due to chemical reactions. In a spatially resolved model this can only be considered as an effect of convection. As densities of the species contained in a mixture in drying applications are only identical in exceptional cases, such convective effects have to be considered in model development. Nevertheless, there exists a couple of drying models in literature, which do not account for convection (e.g. Sloth et al. 2006). Applied to a mixture containing components of different densities, these models violate the continuity equation and are not mass-conservative. The drying models of Seydel (2005), Handscomb, Kraft, and Bayly (2009, and subsequent publications) and Czaputa and Brenn (2012) contain a diffusion driven convective term. Still, their approaches are limited to binary mixtures and Fickian diffusion and do not account for density changing chemical reactions.

In distributed polymerisation systems, convective and diffusive transport involves the polymer species, as well. At first glance, it might be counterintuitive that very large macromolecules diffuse at all. However, this phenomenon is a direct consequence of diffusive transport of the other species in a mixture. If in a binary system a solvent is diffusing into a polymer, diffusion of the polymer component is simply the countermotion to diffusive solvent transport or, from a molecular point of view, solvent and polymer molecules swap places. In mathematical perception, the sum of all diffusive fluxes has to be zero, which implies that there must be a diffusive polymer flux opposite to the solvent flux. Swelling of a polymer as a result of solvent diffusion is an example.

In order to quantify the effect of diffusion and chemical reactions on the velocity field, an additional equation is necessary, as the continuity equation cannot be solved for the velocity field. Typically, the pressure is computed by an additional algebraic equation in CFD methods and then inserted into the momentum balance in order to derive the velocity change. Yet, this course of action is not necessary for both diffusion and reaction induced convection. The volume fraction φ of a single component can be calculated as follows:

$$\varphi_j = \frac{\rho_j}{\rho_j^0} = \frac{c_j MW_j}{\rho_j^0}. \tag{3.1}$$

ρ_j^0 is the reference density of j, which is - under the assumption that there are no mixing effects on the partial volumes of the species, in other words no excess volumes occur - the density of the pure substance j and constant. The sum of all volume fractions is naturally one, which gives a closing condition

$$\sum \varphi_j = \sum \frac{\rho_j}{\rho_j^0} = \sum \frac{c_j MW_j}{\rho_j^0} = 1. \tag{3.2}$$

Having n components, only $n-1$ partial densities or concentrations are independent. A velocity field must ensure, that equation 3.2 is always fulfilled. This closing condition can, hence, be applied at boundaries, where all fluxes are balanced in algebraic equations. A constraint for the component transport equations is derived by taking the derivative of equation 3.2 with respect to time.

In the vast majority of drying processes with or without chemical reactions the assumption of constant densities ρ_j^0 and molar weights is valid and will therefore be applied in the following. An approach for cases, where excess volumes cannot be neglected, will subsequently be discussed in short. In polymerisation processes the averaged molar weight of the polymer species is typically not constant any more, but a function of both time and space. This peculiarity will be adresses in section 3.1.3.

3.1.1 Constant Physical Properties

The derivative of equation 3.2 with respect to time is under the assumptions of $\rho_j^0 = const.$ and $MW_j = const.$

$$\sum \frac{1}{\rho_j^0} \frac{\partial \rho_j}{\partial t} = \sum \frac{MW_j}{\rho_j^0} \frac{\partial c_j}{\partial t} = 0. \tag{3.3}$$

Inserting the transport equation 2.11 and using again the closing condition 3.2 one can obtain a relation between diffusion, chemical reactions and convection

$$0 = \sum \frac{1}{\rho_j^0} \left(-\nabla \left(\rho_j \vec{v} + \vec{j}_j \right) + r_j^F MW_j \right)$$

$$\nabla \left(\sum \frac{\rho_j}{\rho_j^0} \vec{v} \right) = -\nabla \sum \frac{\vec{j}_j}{\rho_j^0} + \sum \frac{r_j^F MW_j}{\rho_j^0}$$

$$\nabla \vec{v} = -\nabla \sum \frac{\vec{j}_j}{\rho_j^0} + \sum \frac{r_j^F MW_j}{\rho_j^0}. \tag{3.4}$$

In a molar frame of reference, using the second part of equation 3.3 and equation 2.29, in a similar manner

$$\nabla \vec{v}^N = -\nabla \sum \vec{J}_j^N \frac{MW_j}{\rho_j^0} - \sum \frac{r_j^F MW_j}{\rho_j^0}. \tag{3.5}$$

is derived. Without chemical reactions, integration of equation 3.4 leads to

$$\vec{v} = \vec{v}^0 - \sum \frac{\vec{j}_j}{\rho_j^0}. \tag{3.6}$$

The integration constant $\vec{v}^0$ is the bulk velocity, which is independent from diffusion and chemical reactions and contains the divergence-free part of the velocity field. Additionally to this bulk velocity, diffusion and chemical reactions can be considered as two independent, seperate contributions. Whereas the diffusion induced velocity $\vec{v}^D$ or $\vec{v}^{D,N}$ can be derived by integration of the first term of equations 3.4 or 3.5

$$\vec{v}^D = -\sum \frac{\vec{j}_j}{\rho_j^0} \qquad \vec{v}^{D,N} = -\sum \vec{J}_j^N \frac{MW_j}{\rho_j^0}, \qquad (3.7)$$

the reaction induced velocity contribution $\vec{v}^R$ cannot be calculated directly. With

$$\nabla \vec{v}^R = \sum \frac{r_j^F MW_j}{\rho_j^0} \qquad (3.8)$$

a first order equation is obtained, which, generally, is underconstrained with respect to all elements of $\vec{v}^R$. It can indeed be used in a flow solver for incompressible liquids instead of the typical velocity constraint 2.8, which is to be modified in this case as $\rho \neq const$. Still, this requires a solution of the momentum balance in order to obtain the velocity field, even if one is solely interested in component transport due to chemical reactions in absence of bulk convection. Nevertheless, for the special cases of one-dimensional problems equation 3.8 simply is a boundary value problem, which can be solved for $\vec{v}^R$ with one, single boundary condition.

Notably, the reaction induced velocity $\vec{v}^R$ is identical in both mass and molar based notations as can be seen from the last terms of equations 3.4 and 3.5. Whereas this fact might be surprising in the first place, it becomes clear if a volume changing reaction within a batch reactor is considered. The fluid level inside the reactor is raised or lowered according to the volume change with its alteration being identical to the average reaction induced velocity at the liquid surface. Of course, this effect will take place in the same way regardless whether a molar or a mass based formulation of the problem is applied. As the bulk velocity is independent from the choice of the frame of reference as well, mass and molar based formulations only differ in regard of diffusion and the velocity contribution related to this effect.

Special Case: (Pseudo-)Binary Fickian Diffusion

Inserting Fickian diffusion according to equation 2.51 into the diffusion induced velocity correlation 3.7 yields

$$\vec{v}^D = -\sum \frac{\vec{j}_j}{\rho_j^0} = -\sum \frac{1}{\rho_j^0} D\left(-\nabla\rho_j + w_j\nabla\rho\right)$$

$$= D\nabla\sum \frac{\rho_j}{\rho_j^0} - D\frac{1}{\rho}\nabla\rho\sum \frac{\rho_j}{\rho_j^0} = -\frac{1}{\rho}D\nabla\rho = -D\nabla\ln\rho. \qquad (3.9)$$

The total flux and the transport equation of a component j then become with $\vec{v} = \vec{v}^0 + \vec{v}^D + \vec{v}^R$

$$\rho_j\vec{v} + \vec{j}_j = \rho_j\left(\vec{v}^0 - \frac{1}{\rho}D\nabla\rho + \vec{v}^R\right) - D\nabla\rho_j + w_j\nabla\rho$$

$$= \rho_j\left(\vec{v}^0 + \vec{v}^R\right) - D\nabla\rho_j \qquad (3.10)$$

$$\frac{\partial\rho_j}{\partial t} = -\nabla\left(\rho_j\left(\vec{v}^0 + \vec{v}^R\right) - D\nabla\rho_j\right) + r_j^F MW_j, \qquad (3.11)$$

so that that the diffusion related velocity contribution can be eliminated. In a molar based notation the same simplification can be applied, so that the diffusion related velocity part and the transport equation become

$$\vec{v}^{D,N} = -\frac{1}{c}D\nabla c = -D\nabla\ln c \qquad (3.12)$$

$$\frac{\partial c_j}{\partial t} = -\nabla\left(c_j\left(\vec{v}^0 + \vec{v}^R\right) - D\nabla c_j\right) + r_j^F. \qquad (3.13)$$

As the remaining velocity components are independent from a mass or molar notation, equations 3.11 and 3.13 are identical and can be simply transformed into one another by multiplication with the (constant) molar mass MW_j. In absence of a bulk flow and chemical reactions they become Fick's second law. A purely diffusive problem may exhibit a very large convective flux, if the specific volumes ($\frac{1}{\rho_j^0}$, $\frac{MW_j}{\rho_j^0}$) of the diffusing species differ strongly. From a numerical point of view, this may introduce oscillations in higher order schemes so that elimination of the convective flux as in equation 3.13 can be advantageous.

3.1.2   Consideration of Mixture Effects

If the volume occupied by a certain mass of a component j depends on the other components in the mixture, the corresponding density ρ_j^0 is not constant anymore, but a function of the mixture composition, as well. The derivative of equation 3.2 is in this case

$$\sum_j \frac{\partial}{\partial t}\left(\frac{\rho_j}{\rho_j^0}\right) = \sum_j \left(\frac{1}{\rho_j^0}\frac{\partial \rho_j}{\partial t} - \frac{\rho_j}{\rho_j^{02}}\frac{\partial \rho_j^0}{\partial t}\right) = 0. \tag{3.14}$$

Assuming that densities depend on the mass fractions of the components in the mixture ($\rho_j^0 = f(w_k)$) and that corresponding partial derivatives $\frac{\partial \rho_j^0}{\partial w_k}$ are known,

$$\sum_j \left(\frac{1}{\rho_j^0}\frac{\partial \rho_j}{\partial t} - \frac{\rho_j}{\rho_j^{02}}\sum_k \frac{\partial \rho_j^0}{\partial w_k}\frac{\partial w_k}{\partial t}\right) = 0 \tag{3.15}$$

provides an expression, in which the component balance equations can be inserted, again. The resulting equation can be rather complicated and a straightforward solution for the diffusion and reaction induced velocity contributions is hardly possible. Still, this equation can be used as an additional constraint in the differential-algebraic system of equations in order to solve for the velocity as a seperate variable. This means, that the equations are solved numerically for the overall velocity $\vec{v}$, whereas the single contributions $\vec{v}^D$ and $\vec{v}^R$ remain unknown. Approaches, in which the density ρ_j^0 depends on other quantities such as partial densities, volume fractions etc., would be considered likewise.

3.1.3 Diffusion and Reaction Driven Convection at Variable Molar Weights

In a polymeric system the (averaged) molar mass of the polymer may be a function of time and space. The total derivative of the molar formulation in the closing condition equation 3.2 is

$$\sum \frac{\partial}{\partial t}\frac{c_j MW_j}{\rho_j^0} = \sum \frac{\partial c_j}{\partial t}\frac{MW_j}{\rho_j^0} + \sum \frac{c_j}{\rho_j^0}\frac{\partial MW_j}{\partial t} = 0. \tag{3.16}$$

$$\frac{\partial MW_j}{\partial t} = \frac{\partial(\rho_j/c_j)}{\partial t} = \frac{1}{c_j}\frac{\partial \rho_j}{\partial t} - \frac{\rho_j}{c_j^2}\frac{\partial c_j}{\partial t}$$

The temporal derivative of MW_j can be expressed using equations 2.46 and 2.29:

$$\frac{\partial MW_j}{\partial t} = \frac{1}{c_j}\left(-\nabla\left(c_j MW_j \vec{v}^N + MW_j \vec{J}_j^N\right) + r_j^F MW_j^{inst}\right)$$

$$-\frac{MW_j}{c_j}\left(-\nabla\left(c_j \vec{v}^N + \vec{J}_j^N\right) + r_j^F\right) \tag{3.17}$$

$$\frac{\partial MW_j}{\partial t} = -\underbrace{\left(\vec{v}^N + \frac{\vec{J}_j^N}{c_j}\right)}_{\vec{v}_j}\nabla MW_j + \frac{r_j^F}{c_j}\left(MW_j^{inst} - MW_j\right). \tag{3.18}$$

Equation 3.18 can be considered as a transport equation for the molar weight of a component j, which is advected with the component velocity $\vec{v}_j$. It is necessary to distinguish between the molar weight with regard to the currently produced/consumed molecules MW_j^{inst} in the reaction term and the average molar weight of the yet existing molecules $MW_j = \frac{\rho_j}{c_j}$ which experience advection and diffusion. Inserting equation 3.18 into the closing condition 3.16 gives

$$-\sum \frac{MW_j}{\rho_j^0}\nabla\left(c_j \vec{v}^N + \vec{J}_j^N\right) + \sum r_j^F \frac{MW_j}{\rho_j^0}$$

$$-\sum \frac{c_j}{\rho_j^0}\left(\vec{v}^N + \frac{\vec{J}_j^N}{c_j}\right)\nabla MW_j + \sum r_j^F \frac{MW_j^{inst} - MW_j}{\rho_j^0} = 0$$

$$-\nabla\left(\sum c_j \frac{MW_j}{\rho_j^0}\vec{v}^N + \sum \frac{MW_j}{\rho_j^0}\vec{J}_j^N\right) + \sum \frac{c_j \vec{v}^N + \vec{J}_j^N}{\rho_j^0}\nabla MW_j$$

$$-\sum \frac{c_j}{\rho_j^0}\left(\vec{v}^N + \frac{\vec{J}_j^N}{c_j}\right)\nabla MW_j + \sum r_j^F \frac{MW_j^{inst}}{\rho_j^0} = 0$$

$$-\nabla \vec{v}^N - \nabla \sum \frac{MW_j}{\rho_j^0}\vec{J}_j^N + \sum r_j^F \frac{MW_j^{inst}}{\rho_j^0} = 0. \tag{3.19}$$

The diffusion and reaction induced velocity contributions

$$\vec{v}^{D,N} = -\sum \frac{MW_j}{\rho_j^0}\vec{J}_j^N \tag{3.20}$$

$$\nabla \vec{v}^{m,R} = \sum r_j^F \frac{MW_j^{inst}}{\rho_j^0} \tag{3.21}$$

are nearly identical to the case of constant molar weights. However, the reaction induced velocity relies upon the molar mass of the currently produced molecules MW_j^{inst}, whereas the diffusion driven part is related to the average molar weight of each component.

For obvious reasons, both the concentrations and the partial densities or corresponding values like the first moment need to be evaluated for species with varying molar mass. The easiest way is to use equation 2.46 in order to obtain partial densities using molar fluxes. If the "ordinary" transport equation 2.11 is employed, mass-based diffusive fluxes will have to be obtained from their molar counterpart by the conversion law 2.42.

Special Case: (Pseudo-)binary Fickian Diffusion

The general expression of diffusion driven convection for variable molar masses is akin to the one of constant values. The diffusion induced velocity in case of (pseudo) binary Fickian diffusion is however different:

$$\vec{v}^{D,N} = -\sum \frac{MW_j}{\rho_j^0}\left(-D\nabla c_j + D\frac{c_j}{c}\nabla c\right)$$

$$\vec{v}^{D,N} = +D\nabla \sum c_j \frac{MW_j}{\rho_j^0} - D\sum \frac{c_j}{\rho_j^0}\nabla MW_j - D\frac{1}{c}\nabla c$$

$$\vec{v}^{D,N} = -D\sum \frac{c_j}{\rho_j^0}\nabla MW_j - D\frac{1}{c}\nabla c. \tag{3.22}$$

The second term is the diffusion driven velocity 3.12 in case of constant molar weights, whereas the first term accounts for the spatial variation of molar masses. The transport equation then becomes

$$\frac{\partial c_j}{\partial t} = -\nabla\left(c_j\left(\vec{v}^0 + \vec{v}^R - D\sum_k \frac{c_k}{\rho_k^0}\nabla MW_k\right) - D\nabla c_j\right) + r_j^F. \tag{3.23}$$

This expression can be transformed into a mass based notation using equations 2.44 and 2.45 and the component velocity $\vec{v}_j$:

$$\frac{\partial \rho_j}{\partial t} = -\nabla(\rho_j \vec{v}_j) + r_j^F MW_j \tag{3.24}$$

$$\frac{\partial \rho_j}{\partial t} = -\nabla\left(\rho_j\left(\vec{v}^0 + \vec{v}^R - D\sum_k \frac{c_k}{\rho_k^0}\nabla MW_k\right) - \frac{\rho_j}{c_j}D\nabla c_j\right). \tag{3.25}$$

Notably, the diffusive flux does not depend on gradients of the component's partial density, but of the corresponding concentration. For species with a varying molar weight, equation 2.57 can therefore no longer be used in order to calculate diffusive mass fluxes. Neither the fluxes $\vec{j}_j = -\rho D \nabla x_j$ nor the diffusion induced mass-averaged velocity $\vec{v}^D = -\sum_j \frac{\vec{j}_j}{\rho_j^0}$ refer to the molar weights of diffusing species so that the effect of varying average molar weight of a polymer species will not be modelled correctly (balance equation 3.11 assuming constant molar mass will be obtained). If one is explicitly interested in the diffusive mass fluxes $\vec{j}_j$ and the diffusion driven velocity $\vec{v}^D$, these can be obtained by the conversion law 2.42 and either equation 2.35 or 3.7

$$\vec{j}_j = -\rho D \nabla w_j + w_j \rho D \frac{\nabla MW_j}{MW_j} - w_j D \sum_k \rho_k \frac{\nabla MW_k}{MW_k} \qquad (3.26)$$

$$\vec{v}^D = -D \frac{\nabla \rho}{\rho} - D \sum_k \frac{\rho_k}{\rho_k^0} \frac{\nabla MW_k}{MW_k} + \frac{1}{\rho} D \sum_k \rho_k \frac{\nabla MW_k}{MW_k}. \qquad (3.27)$$

3.1.4 Transport of Polymer - Quasi-Steady-State Assumption

The quasi-steady-state assumption implicitly contains premises regarding the transport behaviour of polymer radicals. The total concentration of living chains $[R_{tot}]$ is an instantaneous value, which is calculated by an algebraic expression solely depending on the local concentrations of educts. Hence, transport of living chains is presumed not to play a role. Due to the limited lifetime of a polymer radical, which amounts to fractions of seconds (Hutchinson 2005, p. 160), this assumption is typically very reasonable. Before living chains may diffuse and slowly exchange places with other molecules, they will rather be terminated. Convection and diffusion will affect the educt concentrations and thus the algebraically determined polymer values. Moreover, the concentration and partial density of the (dead) polymer are cumulative values and may experience significant changes due to transport. Generally, the balance equations of the polymer concentration and the partial density can be expressed by (equations 2.29, 2.46)

$$\frac{\partial [P]}{\partial t} = -\nabla \left([P] \vec{v}^N + \vec{J}_P^N \right) + r_P^F \qquad (3.28)$$

$$\frac{\partial \rho_P}{\partial t} = -\nabla \left(\rho_P \vec{v}^N + MW_P \vec{J}_P^N \right) + r_P^F MW_P^{inst}. \qquad (3.29)$$

The reaction term must be related to the instantaneous molar weight of the currently produced polymer MW_P^{inst}, whereas diffusion affects all polymer molecules with the average molar weight $MW_P = \frac{\rho_P}{[P]}$. The different velocity contributions have to be calculated as derived in section 3.1.3. This approach is valid for both Maxwell-Stefan and Fickian diffusion. In case of (pseudo-) binary Fickian diffusion the corresponding simplified equations 3.23 and 3.25 can be used

$$\frac{\partial [P]}{\partial t} = -\nabla \left([P] \left(\vec{v}^0 + \vec{v}^R - D\frac{[P]}{\rho_P^0}\nabla\frac{\rho_P}{[P]} \right) - D\nabla [P] \right) + r_P^F \tag{3.30}$$

$$\frac{\partial \rho_P}{\partial t} = -\nabla \left(\rho_P \left(\vec{v}^0 + \vec{v}^R - D\frac{[P]}{\rho_P^0}\nabla\frac{\rho_P}{[P]} \right) - \frac{\rho_P}{[P]}D\nabla [P] \right)$$
$$+ r_P^F MW_P^{inst}. \tag{3.31}$$

Diffusive transport of the polymer's partial density is only affected by the concentration of polymer, not its partial density or mass fraction. From this it follows that a gradient in the partial density of the polymer will not lead to a diffusive flux as long as the polymer concentration does not vary, as well. This means, that variations in the polymer's molar weight will be preserved as long as there are no modifications due to reactions. If, in contrast, the velocity contribution due to variations in the molar weights was neglected, as in Fick's second law (equations 3.13 and 3.11)

$$\frac{\partial [P]}{\partial t} = -\nabla \left([P] \left(\vec{v}^0 + \vec{v}^R \right) - D\nabla [P] \right) + r_P^F$$

$$\frac{\partial \rho_P}{\partial t} = -\nabla \left(\rho_P \left(\vec{v}^0 + \vec{v}^R \right) - D\nabla\rho_P \right) + r_P^F MW_P^{inst},$$

diffusion would be calculated independently for the polymer concentration and partial density. Spatial variations in the polymer's partial density would flatten out, even if the concentration of polymer was constant. As a result, a gradient in the molar mass would always vanish unphysically after a sufficiently long time. In a molecular sense, Fick's second law, applied to both the concentration and partial density of the polymer, means that the polymer molecules themselves change places with each other, for example a chain consisting of one thousand monomer units with a chain of length five hundred. This is not only counterintuitive, but also contrary to well-known observations like the gel-effect, where for a growing polymer concentration the polymer molecules become more or less fixed and termination reactions due to encounter of two radical chains are scarcely taking place any more. Hence, Fick's second law cannot be applied to diffusion of polymer of variable molar weight without the above-mentioned modifications.

3.1.5 Transport of Statistical Moments

If the method of moments is applied to a distributed system, partial differential equations for the statistical moments of the distributions of both living and dead chains need to be evaluated. As has been stated above, the lifetime of polymer radicals is very short. Transport of living chains should therefore be negligible. Yet, a similar consideration of all polymer values in the mixture appeared to improve numerical stability in simulations and the balance equations of living and dead chain's moments are mathematically equivalent. Thus, the same transport approach will be applied to both sets of equations in the following.

Starting with the balance equation of dead chains of length s in terms of concentrations (equation 2.29), multiplication with s^k and subsequent summation over all chain lengths s yields

$$\sum_s s^k \frac{\partial [P_s]}{\partial t} = -\sum_s s^k \nabla \left([P_s] \vec{v}^N + \vec{J}_{P_s}^N \right) + \sum_s s^k r_{P_s}^F$$

$$\frac{\partial \sum_s s^k [P_s]}{\partial t} = -\nabla \left(\sum_s s^k [P_s] \vec{v}^N + \sum_s s^k \vec{J}_{P_s}^N \right) + \sum_s s^k r_{P_s}^F$$

$$\frac{\partial \zeta_k}{\partial t} = -\nabla \left(\zeta_k \vec{v}^N + \vec{J}_{\zeta_k}^N \right) + r_{\zeta_k}^F. \tag{3.32}$$

The s^k terms and summations are not depending on time and space and can be shifted into the differential operators, where the concentrations in the accumulation and convection terms can easily be replaced by the moments or the respective rates of formation. The transformation of the reaction term corresponds to the classical approach in the method of moments, which has been introduced in section 2.3.3. In contrast, the diffusive behaviour of the moments has merely been condensed into a flux $\vec{J}_{\zeta_k}^N = \sum_s s^k \vec{J}_{P_s}^N$, which is yet unspecified and needs further consideration.

A naive description of moment diffusion follows Fick's second law 3.13:

$$\frac{\partial P_s}{\partial t} = -\nabla \left([P_s] \left(\vec{v}^0 + \vec{v}^R \right) - D\nabla [P_s] \right) + r_{P_s}^F \tag{3.33}$$

$$\sum_s s^k \frac{\partial P_s}{\partial t} = -\nabla \left(\sum_s s^k [P_s] \left(\vec{v}^0 + \vec{v}^R \right) - D\nabla \sum_s s^k [P_s] \right) + \sum_s s^k r_{P_s}^F \tag{3.34}$$

$$\frac{\partial \zeta_k}{\partial t} = -\nabla \left(\zeta_k \left(\vec{v}^0 + \vec{v}^R \right) - D\nabla \zeta_k \right) + r_{\zeta_k}^F. \tag{3.35}$$

In this way a transport equation for the statistical moments is obtained, in which moments are treated like normal species with Fickian diffusion. Furthermore, the moment equations are closed without the need of an additional condition. This approach for moment diffusion has - without derivation - already been proposed by Arriola (1989, 317f.). The diffusive flux of a moment ζ_k is then

$$\vec{J}^N_{\zeta_k} = -D\nabla\zeta_k + D\frac{\zeta_k}{c}\nabla c. \qquad (3.36)$$

Wheras this way of proceeding follows a straightforward derivation, it exhibits the same shortcomings which have been discussed before regarding the application of Fick's second law when using the QSSA. The underlying equation 3.33 assumes Fickian diffusion of polymer molecules of chain length s solely depending on their concentration $[P_s]$, regardless of the total polymer concentration. Polymer chains of different length will diffuse against one another, if their gradients have opposite directions. Again, given enough time, this implies vanishing gradients of the concentrations of all single chain lengths and, hence, of the molar mass of the polymer.

Generally, the zeroth and first moments are equivalent to the total polymer concentration and partial density, respectively. In terms of the distribution of dead chains, with $\rho_P = \zeta_1 MW_M$, $MW_P = \frac{\zeta_1}{\zeta_0}MW_M$ and equation 2.46 the transport equations without reactions can be written as

$$\frac{\partial \zeta_0}{\partial t} = \frac{\partial [P]}{\partial t} = -\nabla\left(\zeta_0\vec{v}^N + \vec{J}^N_{\zeta_0}\right) = -\nabla\left(\zeta_0\vec{v}^N_{\zeta_0}\right) \qquad (3.37)$$

$$\frac{\partial \zeta_1}{\partial t} = \frac{1}{MW_M}\frac{\partial \rho_P}{\partial t} = -\nabla\left(\zeta_1\vec{v}^N + \frac{\zeta_1}{\zeta_0}\vec{J}^N_{\zeta_0}\right) = -\nabla\left(\zeta_1\vec{v}^N_{\zeta_0}\right) \qquad (3.38)$$

These transport expression are equivalent to equations 3.30 and 3.31 from the QSSA. The diffusive flux of the first moment is then depending on the gradient of the zeroth moment

$$\vec{J}^N_{\zeta_1} = \frac{\zeta_1}{\zeta_k}\vec{J}^N_{\zeta_k}. \qquad (3.39)$$

A generalisation of this relationship to higher moments can be accomplished by the Maxwell-Stefan equations, if the following assumptions for the binary diffusion coefficients between polymer chains P_s and bulk (B) molecules u and v of different kind are made:

polymer $\leftrightarrow$ bulk	$Đ_{P_s u} = Đ_{Pu}$	
polymer $\leftrightarrow$ polymer	$Đ_{P_s P_t} = Đ_{PP}$,	$Đ_{PP} \to 0$
bulk $\leftrightarrow$ bulk	$Đ_{uv}$	

The first presumption implies, that the diffusive interaction between polymer and other molecules does not depend on the polymer's chain length and is equal for all polymer chains. This simplification is certainly not true for very short polymer chains (a radical of length one virtually is a monomer molecule). Nevertheless, their concentration is very low, so that it is common practice to neglect the peculiar behaviour of short chains in polymer reaction engineering without introducing significant errors (cmp. section 2.3.1). The binary diffusion coefficient $Đ_{Pu}$ may, still, be different for various kinds of bulk molecules u.

Diffusion between polymer chains is also expected to be independent from the chain lengths. Moreover, it is assumed, that there is virtually no diffusion of polymer molecules against themselves with the binary diffusion coefficient $Đ_{PP}$ tending to zero.

The interrelation between the driving force of a component and the diffusive fluxes, equation 2.60, can be expressed in terms of the bulk components u, v and polymer chains P_s

$$c\vec{d}_u = \sum_k \frac{x_u \vec{J}_k^N - x_k \vec{J}_u^N}{Đ_{uk}}$$

$$= x_u \left(\sum_s \frac{\vec{J}_{P_s}^N}{Đ_{Pu}} + \sum_{v \in B} \frac{\vec{J}_v^N}{Đ_{uv}} \right) - \vec{J}_u^N \left(\sum_s \frac{x_{P_s}}{Đ_{Pu}} + \sum_{v \in B} \frac{x_v}{Đ_{uv}} \right).$$

The mole fractions and diffusive fluxes of the single polymer chains can be eliminated using $x_P = \sum_s x_{P_s}$ and $\sum_s \vec{J}_{P_s}^N = - \sum_{v \in B} \vec{J}_v^N$:

$$c\vec{d}_u = x_u \sum_{v \in B} \vec{J}_v^N \left(\frac{1}{Đ_{uv}} - \frac{1}{Đ_{Pu}} \right) - \vec{J}_u^N \left(\sum_{v \in B} \frac{x_v}{Đ_{uv}} + \frac{x_P}{Đ_{Pu}} \right). \tag{3.40}$$

Setting $P = n$, this is just the same equation as 2.63, in which the fluxes $\vec{J}_n^N$ of the n-th component have been eliminated. This means that the bulk components are treated the same way as in ordinary Maxwell-Stefan diffusion. Their diffusive fluxes $\vec{J}_u^N$ just depend on the driving forces of the bulk species $\vec{d}_u$, whereas the driving forces of the polymer chains do not affect diffusive bulk behaviour. The driving force of a polymer species of chain length s is

$$c\vec{d}_{P_s} = \sum_k \frac{x_{P_s}\vec{J}_k^N - x_k\vec{J}_{P_s}^N}{Ð_{Pk}}$$

$$= x_{P_s}\left(\sum_t \frac{\vec{J}_{P_t}^N}{Ð_{PP}} + \sum_{v \in B} \frac{\vec{J}_v^N}{Ð_{Pv}}\right) - \vec{J}_{P_s}^N\left(\sum_t \frac{x_{P_t}}{Ð_{PP}} + \sum_{v \in B} \frac{x_v}{Ð_{Pv}}\right).$$

Multiplication with $Ð_{PP}$ gives

$$Ð_{PP}c\vec{d}_{P_s} = x_{P_s}\sum_t \vec{J}_{P_t}^N + x_{P_s}\sum_{v \in B}\frac{Ð_{PP}}{Ð_{Pv}}\vec{J}_v^N - \sum_t x_{P_t}\vec{J}_{P_s}^N - \sum_{v \in B}\frac{Ð_{PP}}{Ð_{Pv}}x_v\vec{J}_{P_s}^N.$$

The limit value for $Ð_{PP} \to 0$ yields

$$\vec{J}_{P_s}^N = \frac{x_{P_s}}{\sum_t x_{P_t}}\sum_t \vec{J}_{P_t}^N = \frac{[P_s]}{[P]}\vec{J}_P^N, \tag{3.41}$$

where the diffusive flux of the polymer component as a whole $\vec{J}_P^N$ can be calculated from the bulk components' diffusion

$$\vec{J}_P^N = \sum_s \vec{J}_{P_s}^N = -\sum_{v \in B}\vec{J}_v^N. \tag{3.42}$$

The diffusive flux of a moment ζ_k is thus

$$\vec{J}_{\zeta_k}^N = \sum_s s^k \vec{J}_{P_s}^N = \sum_s s^k \frac{[P_s]}{[P]}\vec{J}_P^N$$

$$\vec{J}_{\zeta_k}^N = \frac{\zeta_k}{\zeta_0}\vec{J}_P^N. \tag{3.43}$$

Living chains exhibit a short livetime in comparison to the diffusive time scale. Their transport behaviour hence hardly affects chemical reactions. Regard of their moments' diffusion can improve numerical stability, yet. As, in terms of diffusion, living and dead chains cannot be distinguished, the diffusive fluxes are

$$\vec{J}_{\zeta_k}^N = \frac{\zeta_k}{\lambda_0 + \zeta_0}\vec{J}_P^N, \quad \vec{J}_{\lambda_k}^N = \frac{\lambda_k}{\lambda_0 + \zeta_0}\vec{J}_P^N, \quad \vec{J}_P^N = \sum_s \vec{J}_{P_s}^N + \vec{J}_{R_s}^N = -\sum_{v \in B}\vec{J}_v^N. \tag{3.44}$$

Using the bulk moments μ_k, these relations can be expressed as

$$\vec{J}_{\mu_k}^N = \frac{\mu_k}{\mu_0}\vec{J}_P^N, \qquad \vec{J}_{\lambda_k}^N = \frac{\lambda_k}{\mu_0}\vec{J}_P^N. \tag{3.45}$$

General transport equations for the moments are therefore

$$\frac{\partial \zeta_k}{\partial t} = -\nabla \left(\zeta_k \vec{v}^N + \frac{\zeta_k}{\lambda_0 + \zeta_0} \vec{J}_P^N \right) + r_{\zeta_k}^F = -\nabla \left(\zeta_k \vec{v}_P^N \right) + r_{\zeta_k}^F \qquad (3.46)$$

$$\frac{\partial \lambda_k}{\partial t} = -\nabla \left(\lambda_k \vec{v}^N + \frac{\lambda_k}{\lambda_0 + \zeta_0} \vec{J}_P^N \right) + r_{\lambda_k}^F = -\nabla \left(\lambda_k \vec{v}_P^N \right) + r_{\lambda_k}^F. \qquad (3.47)$$

As the component velocity of all moments is equal ($\vec{v}_P = \vec{v}^N + \frac{\vec{J}_P^N}{\lambda_0 + \zeta_0}$), transport will advect profiles in the chain length distribution but not smear out differences. For (pseudo-)binary Fickian diffusion, $\vec{v}^{D,N}$ is according to equation 3.22

$$\vec{v}^{D,N} = -\frac{1}{c}D\nabla c - MW_M \frac{\lambda_0 + \zeta_0}{\rho_P^0} D\nabla \frac{\lambda_1 + \zeta_1}{\lambda_0 + \zeta_0} \qquad (3.48)$$

and the transport equations can be expressed in terms of Fick's second law as

$$\frac{\partial *}{\partial t} = -\nabla \left(* \left(\vec{v}^0 + \vec{v}^R - MW_M \frac{\lambda_0 + \zeta_0}{\rho_P^0} D\nabla \frac{\lambda_1 + \zeta_1}{\lambda_0 + \zeta_0} \right) \right.$$

$$\left. -\frac{*}{\lambda_0 + \zeta_0} D\nabla (\lambda_0 + \zeta_0) \right) + r_*^F, \qquad * = \zeta_k \text{ or } \lambda_k. \qquad (3.49)$$

3.2 Lumped Modelling - 0D approach

The most simple approximation of droplet drying with or without chemical re-actions is to consider a droplet as one single, ideally mixed volume. As a result, one obtains a very simple system of ordinary differential equations, which de-scribes the evolution of the different mixture components without the effect of transport inside the droplet. Depending on the species and drying conditions this may introduce rather extensive simplifications. The mass transfer Biot number indicates that concentration gradients may be present throughout the process (section 2.5.4). Simulation results can therefore differ from the realistic be-haviour. Still, it is worthwhile to attend to a 0D consideration, first. These kinds of models are easy to derive and implement and they can be evaluated in a very short time. Therefore, lumped models offer quick approximations of reactive drying processes. Moreover, due to the low computational overhead, they are a reasonable option for integrated models, in which a whole spray dryer is con-sidered and a large number of droplets inside a complicated, three-dimensional flow field are simulated. A spatially resolved consideration of transport within the single droplets would be too costly in this case.

3.2.1 General Equations for Reactive Spray Drying

From the reaction engineering point of view, a lumped model resembles a batch execution with the particular difference of an outward flow of volatile species. Balance equations are therefore like in a stirred tank reactor with no inflow and selective outflow. Droplet volume V^d and radius R evolve over time with

$$\frac{dV^d}{dt} = V^d \sum_j r_j^F \frac{MW_j^{inst}}{\rho_j^0} - 4\pi R^2 \sum_j \Omega_j^N \frac{MW_j}{\rho_j^0} \tag{3.50}$$

$$\frac{dR}{dt} = \left(\frac{dV^d}{dR}\right)^{-1} \frac{dV^d}{dt} = \frac{R}{3} \sum_j r_j^F \frac{MW_j^{inst}}{\rho_j^0} - \sum_j \Omega_j^N \frac{MW_j}{\rho_j^0}. \tag{3.51}$$

The first term corresponds to volume changes by chemical reaction, whereas the second term denotes the mass loss due to evaporative molar fluxes Ω_j^N. Again, the molar mass of currently created/consumed molecules MW_j^{inst} has to be considered in the reaction term, whereas vaporisation concerns the averaged molar mass of component j in the mixture. In practice, this distinction only affects polymer molecules, which do not evaporate anyway, and molar weights are constant otherwise. The concentration of a component j varies over time with

$$\frac{dc_j}{dt} = \frac{d\left(N_j/V^d\right)}{dt} = \frac{1}{V^d} \frac{dN_j}{dt} - \frac{N_j}{V^{d2}} \frac{dV^d}{dt}$$

$$\frac{dc_j}{dt} = r_j^F - \frac{3}{R}\Omega_j^N - c_j\frac{3}{R}\frac{dR}{dt}. \tag{3.52}$$

The system of equations is similar, based on mass fluxes and partial densities

$$\frac{dV^d}{dt} = \sum_j V^d r_j^F \frac{MW_j^{inst}}{\rho_j^0} - 4\pi R^2 \sum_j \frac{\Omega_j}{\rho_j^0} \tag{3.53}$$

$$\frac{dR}{dt} = \frac{R}{3} \sum_j r_j^F \frac{MW_j^{inst}}{\rho_j^0} - \sum_j \frac{\Omega_j}{\rho_j^0} \tag{3.54}$$

$$\frac{d\rho_j}{dt} = r_j^F MW_j^{inst} - \frac{3}{R}\Omega_j - \rho_j\frac{3}{R}\frac{dR}{dt}. \tag{3.55}$$

The energy balance needs to consider the heat release by chemical reactions, heat transfer across the interface to the drying gas and cooling due to evaporation. Assuming constant heat capacities $c_{p,j}$, the temperature evolves with

$$\frac{\partial T}{\partial t} = \frac{1}{\rho \bar{c}_p}\left(\sum_i r_i \Delta h_{R,i} - \frac{3}{R}\left(\dot{q}^\Gamma + \sum_j \Omega_j \Delta h_{v,j}\right)\right) \tag{3.56}$$

3.2.2 Spray Polymerisation - Quasi-Steady-State Assumption

If the QSSA is applied, the total concentration of living chains $[R_{tot}]$ inside the droplet can be calculated according to equation 2.89. The concentrations of solvent (S), monomer (M), initiator (I) and dead polymer chains (D) evolve with

$$\frac{d\,[S]}{dt} = -\frac{3}{R}\Omega_S^N - [S]\frac{3}{R}\frac{dR}{dt} \tag{3.57}$$

$$\frac{d\,[I]}{dt} = -k_d\,[I] - [I]\frac{3}{R}\frac{dR}{dt} \tag{3.58}$$

$$\frac{d\,[M]}{dt} = -k_p\,[M]\,[R_{tot}] - 2fk_i\,[I] - k_{trs}\,[S]\,[R_{tot}] - k_{trm}\,[M]\,[R_{tot}]$$

$$- \frac{3}{R}\Omega_M^N - [M]\frac{3}{R}\frac{dR}{dt}$$

$$\approx -k_p\,[M]\,[R_{tot}] - \frac{3}{R}\Omega_M^N - [M]\frac{3}{R}\frac{dR}{dt} \tag{3.59}$$

$$\frac{d\,[P]}{dt} = (k_{td} + 0.5k_{tc})\,[R_{tot}]^2 + (k_{trm}\,[M] + k_{trs}\,[S])\,[R_{tot}] - [P]\frac{3}{R}\frac{dR}{dt}. \tag{3.60}$$

Additionally, the partial density of the polymer is of interest in order to obtain its average molar weight or the cumulated degree of polymerisation. As the formation of polymer is directly connected to the consumption of monomer and accumulation of living chains is negligible, the reaction term can be related to the monomer reactions

$$\frac{d\rho_P}{dt} = -r_M^F MW_M - \rho_P\frac{3}{R}\frac{dR}{dt} = k_p\,[M]\,[R_{tot}]\,MW_M - \rho_P\frac{3}{R}\frac{dR}{dt}. \tag{3.61}$$

The same relationship can be derived, if the rate of formation of dead polymer and the molar weight of the currently generated dead chains $MW_P^{inst} = DP^{inst}MW_M$ are applied. Using the definition of the degree of polymerisation 2.91 one obtains $r_P^F DP^{inst} MW_M = k_p\,[M]\,[R_{tot}]\,MW_M$. Equally, the change in droplet radius due to chemical reactions is related to the molar weight of the instantaneously created polymer and not to the average molar mass of the whole polymer

$$\left.\frac{dR}{dt}\right|^R = \frac{R}{3}\left(\sum_{j\neq P} r_j^F \frac{MW_j}{\rho_j^0} + r_P^F \frac{MW_P^{inst}}{\rho_P^0}\right)$$

$$= \frac{R}{3}\left(\sum_{j\neq P} r_j^F \frac{MW_j}{\rho_j^0} - r_M^F \frac{MW_M}{\rho_P^0}\right). \tag{3.62}$$

Considering propagation being the dominant volume/density changing reaction, this relation can be further simplified to

$$\left.\frac{dR}{dt}\right|^R \approx \frac{R}{3} r_M^F MW_M \left(\frac{1}{\rho_M^0} - \frac{1}{\rho_P^0}\right). \tag{3.63}$$

This assumption is generally true, as the volumetric effects of initiator decomposition or other polymerisation reactions besides propagation are negligibly small. The volume change does not depend on the polymer chain-length. As the volume occupied by a single monomer unit inside a chain is presupposed as being constant, the conversion of monomer to polymer determines the overall volume change.

3.2.3 Spray Polymerisation - Method of Moments

An implementation of the method of moments into a lumped model is straight-forward. With the statistical moments having the unit of concentrations, they can be inserted directly into equation 3.52 to

$$\frac{d\lambda_k}{dt} = r_{\lambda_k}^F - \lambda_k \frac{3}{R}\frac{dR}{dt} \tag{3.64}$$

$$\frac{d\zeta_k}{dt} = r_{\zeta_k}^F - \zeta_k \frac{3}{R}\frac{dR}{dt}, \tag{3.65}$$

for which the moments' reaction rates can be taken from Table 2.3. The equations of the non-polymer components are identical to those of the quasi-steady-state-assumption. Additional calculation of the polymer's partial density is not necessary, as the molar weight can be computed from the moments. The effect of density changing polymerisation reactions on the droplet radius can be modelled by equation 3.63 with λ_0 replacing $[R_{tot}]$.

3.3 Distributed Modelling - 1D approach

A lumped approximation of the droplet may fall short by considering the droplet as fully mixed and neglecting transport and spatial gradients. In the following, a one-dimensional representation of reactive droplet drying and droplet polymerisation will be derived. The droplet is considered as a solution, consisting of one quasi-homogeneous liquid phase. A decreased solubility of single components might lead to precipitation (as observed by Franke, Moritz, and Pauer 2017) and the formation of additional phases, which will be undesirable in most applications. Such effects are not currently considered in the model, but could be regarded as model events during the numerical solution or in the post-processing based on threshold concentrations.

General transport equations and boundary conditions for reactive drying processes will be derived in the following two sections. The majority of such applications can be treated sufficiently by these equations. Still, spray polymerisation is a special case of reactive droplet drying, where the product properties, particularly the molar mass, are strongly dependent on the educt concentrations and may vary strongly over time and space. The specific treatment of polymer systems will therefore be addressed in sections 3.3.3 and 3.3.4.

3.3.1 General Equations of the Droplet Continuum

For a better readability, vectors will be reduced to their radial component and the subscript r omitted in the following, as only the radial direction needs to be considered in a one-dimensional model. The nabla operator in radial direction is $\nabla_r v_r = \frac{1}{r^2} \frac{\partial (r^2 v_r)}{\partial r}$ for the divergence of a vector $\vec{v}$, whereas the gradient operator is the same as in cartesian coordinates.

Typically, in binary drying models a transport equation for the mass fraction w_j of one species is derived and the remaining component is calculated by the closing condition. In reactive drying processes the kinetics of the chemical reactions depend on concentrations, so that it is more appropriate to balance the concentrations directly or, alternatively, partial densities. In case of (pseudo-)binary Fickian diffusion, it will be shown later that this approach is advantageous even in the case of pure drying without reaction. Transport of the single compontents inside the droplet is then expressed by equations 2.11 and 2.29:

$$\frac{\partial \rho_j}{\partial t} = -\frac{1}{r^2}\frac{\partial}{\partial r}\left(r^2 \rho_j v + r^2 j_j\right) + r_j^F MW_j^{inst} \tag{3.66}$$

$$\frac{\partial c_j}{\partial t} = -\frac{1}{r^2}\frac{\partial}{\partial r}\left(r^2 c_j v^N + r^2 J_j^N\right) + r_j^F. \tag{3.67}$$

Using equations 2.46 and 2.47 the component balance equations is

$$\frac{\partial \rho_j}{\partial t} = -\frac{1}{r^2}\frac{\partial}{\partial r}\left(r^2 \rho_j v^N + r^2 \frac{\rho_j}{c_j} J_j^N\right) + r_j^F MW_j^{inst} \tag{3.68}$$

$$\frac{\partial c_j}{\partial t} = -\frac{1}{r^2}\frac{\partial}{\partial r}\left(r^2 c_j v + r^2 \frac{c_j}{\rho_j} j_j\right) + r_j^F, \tag{3.69}$$

as well. The reaction induced velocity contribution is given by the boundary value problem

$$\frac{1}{r^2}\frac{\partial \left(r^2 v^R\right)}{\partial r} = -\sum \frac{r_j^F MW_j}{\rho_j^0}, \tag{3.70}$$

which is the one-dimensional case of equation 3.8.

Convection inside the droplet is mainly driven by diffusion. The reaction related contribution is typically smaller. Therefore, convective energy transport will not exceed energy transport by diffusion significantly. The Lewis number Le contains the ratio of temperature conduction to diffusive matter transport

$$Le = \frac{a}{D} = \frac{\frac{\lambda}{\rho c_p}}{D} = \frac{O\left(10^{-1}\right)}{O\left(10^3\right) O\left(10^3\right) O\left(10^{-9}\right)} \approx O\left(10^2\right). \tag{3.71}$$

The thermal diffusivity a is two orders of magnitude higher than matter diffusivity, so that energy transport is dominated by heat conduction. Assuming constant heat capacities, the enthalpy balance can be reduced to the temperature balance 2.24, which is very simple in this case

$$\frac{\partial T}{\partial t} = -\frac{\nabla \vec{q}}{\sum \rho_j c_{p_j}} - \frac{\sum r_j^F MW_j h_j}{\sum \rho_j c_{p_j}}, \tag{3.72}$$

as energy transport by convection and diffusion is neglected. As has been discussed in section 2.5.4, the droplet can be assumed as thermally mixed, if the Nusselt number is around 4 or smaller. In this case the chemical reactions can be summed over the droplet radius and inserted into the lumped energy balance equation 3.56.

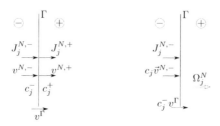

Figure 3.1: Transport across boundaries in a one-dimensional description, left: general case of molar transport, right: transport across the droplet surface.

3.3.2 Boundary Conditions

The boundary conditions at the droplet's centre and at the interface to the drying gas can be derived using equation 2.2

$$\left(\psi^+ \vec{v}^+ + \phi_\psi^+ - \psi^+ \vec{v}^\Gamma - \psi^- \vec{v}^- - \phi_\psi^- + \psi^- \vec{v}^\Gamma \right) \cdot \vec{n} - \sigma_\psi^\Gamma = 0.$$

At both boundaries, there is no production/consumption of conserved quanties so that the source term is zero and only the part inside the brackets remains. Due to symmetry, this formula can be reduced to one-dimension in radial direction

$$\psi^+ v^+ + \phi_\psi^+ - \psi^+ v^\Gamma - \psi^- v^- - \phi_\psi^- + \psi^- v^\Gamma = 0. \tag{3.73}$$

The vector directions and the nomenclature of $+$ and $-$ indices are sketched for transport in a molar average description in Figure 3.1, left frame.

Droplet Centre

The droplet centre is spatially fixed at $r = 0$ so that $v^\Gamma\big|_{r=0} = 0$. Due to symmetry, scalar values on both sides of the interface are identical. Additionally, the slope of any quantity has to be continous over the droplet centre. Both prerequisites can only be fulfilled in case of zero gradients of the respective values:

$$\frac{\partial \rho_j}{\partial r}\bigg|_{r=0} = 0, \qquad \frac{\partial w_j}{\partial r}\bigg|_{r=0} = 0 \tag{3.74}$$

$$\frac{\partial c_j}{\partial r}\bigg|_{r=0} = 0, \qquad \frac{\partial x_j}{\partial r}\bigg|_{r=0} = 0 \tag{3.75}$$

$$\frac{\partial T}{\partial r}\bigg|_{r=0} = 0. \tag{3.76}$$

This implicitly involves zero diffusive fluxes and, due to symmetry and mass conservation, a zero velocity at the droplet's centre. Explicitly, this is derived by the precondition of symmetry that vector quantities on both sides of the interface have the same magnitude, but opposite direction. Setting $\psi = \rho$, with $\rho^-|_{r=0} = \rho^+|_{r=0} = \rho|_{r=0}$ and $v^-|_{r=0} = -v^+|_{r=0} = -v|_{r=0}$, yields

$$\left[\rho^+ v^+ - \rho^- v^-\right]_{r=0} = [2\rho v]_{r=0} = 0$$
$$v|_{r=0} = 0. \tag{3.77}$$

With $\psi = \rho_j$ and $-j_j^-\big|_{r=0} = j_j^+\big|_{r=0} = j_j\big|_{r=0}$ the boundary condition for the diffusive fluxes is

$$\left[j_j^+ - j_j^-\right]_{r=0} = 0$$
$$j_j\big|_{r=0} = 0. \tag{3.78}$$

Due to the absence of diffusive fluxes the diffusion driven velocities v^D and $v^{D,N}$ are zero. As the overall velocity is zero, the reaction induced contribution needs to be zero, as well, which provides the boundary condition for solving equation 3.70:

$$v^R\big|_{r=0} = 0. \tag{3.79}$$

Interface to the Drying Gas

The droplet's boundary to the drying gas at $r = R$ is moving due to evaporation and volume-changing chemical reactions with the velocity v^Γ. Wheras the values inside the droplet ('−'-side) are known or can be calculated, the fluxes in the drying gas are not available in detail. Mass, molar and heat fluxes are calculated according to linear driving forces, where the transfer coefficients are obtained from appropriate dimensionless correlations. In the interface balance equations the terms considering the drying gas (all terms with superscript '+') are subsumed into the linear driving forces Ω, Ω^N or $\dot{q}^\Gamma$, respectively, in which the motion of the interface, the detailed flow regime and the interplay of convective and diffusive fluxes in the drying gas are compressed into one simple expression (see the right frame of Figure 3.1). Mass and molar balances at the droplet boundary are then

$$\left[\rho_j\left(v - v^\Gamma\right) + j_j\right]_{r=R} = \Omega_j \tag{3.80}$$
$$\left[c_j\left(v^N - v^\Gamma\right) + J_j^N\right]_{r=R} = \Omega_j^N. \tag{3.81}$$

For reasons of better readability, the superscript '−' has been dropped for the values on the droplet side. These equations can be expressed in terms of the component velocity v_j

$$\left[\rho_j \left(v_j - v^\Gamma\right)\right]_{r=R} = \Omega_j \tag{3.82}$$

$$\left[c_j \left(v_j - v^\Gamma\right)\right]_{r=R} = \Omega_j^N, \tag{3.83}$$

as well, which can be simplified for non-volatile components to

$$v_j\big|_{r=R} = v^\Gamma. \tag{3.84}$$

The component velocities of non-evaporating species at the interface to the drying gas is therefore identical to the boundary velocity v^Γ. Division of equation 3.80 with ρ^0 and summation over all components j gives

$$\left[\sum_j \frac{\rho_j}{\rho_j^0} \left(v - v^\Gamma\right) + \sum_j \frac{\rho_j}{\rho_j^0} j_j\right]_{r=R} = \sum_j \frac{\Omega_j}{\rho_j^0} \tag{3.85}$$

$$\left[v - v^\Gamma - v^D\right]_{r=R} = \sum_j \frac{\Omega_j}{\rho_j^0}. \tag{3.86}$$

As no convection is imposed, $v = v^R + v^D$ and the interface velocity v^Γ is

$$v^\Gamma = v^R\big|_{r=R} - \sum_j \frac{\Omega_j}{\rho_j^0}. \tag{3.87}$$

The first term expresses interface motion/droplet shrinkage due to chemical reactions. The evaporation term is identical to the lumped model and can be regarded as the evaporation driven interface velocity $v^{\Gamma,evap}$

$$v^{\Gamma,evap} = -\sum_j \frac{\Omega_j}{\rho_j^0} = -\sum_j \frac{MW_j \Omega_j^N}{\rho_j^0}. \tag{3.88}$$

Equation 3.80 then becomes the general boundary condition for a component j

$$\left[\rho_j \left(v^D - v^{\Gamma,evap}\right) + j_j\right]_{r=R} = \Omega_j. \tag{3.89}$$

Similarly, the expression in molar notation can be derived to

$$\left[c_j \left(v^{D,N} - v^{\Gamma,evap}\right) + J_j^N\right]_{r=R} = \Omega_j^N. \tag{3.90}$$

The droplet radius R changes according to the interface velocity

$$\frac{dR}{dt} = \vec{v}^{\Gamma}. \tag{3.91}$$

Obviously, the boundary conditions 3.89 and 3.90 are not directy coupled to chemical reactions. This is due to the fact, that reaction induced transport is of convective nature and therefore a first order phenomenon, which requires only one boundary condition. The constraint for convection and the boundary value problem 3.70 is already set at the droplet centre (equation 3.79). Both phenomena, reaction and diffusion induced convection, are independent effects and may be superimposed. The interplay of evaporation and diffusion at the droplet's boundary is not altered by chemical reactions. Still, concentration profiles are shifted in space due to volume changing reactions and modified due to consumption/generation in the droplet's bulk.

In case of (pseudo-)binary Fickian diffusion with constant molar masses, the equations can be simplified, again. Inserting $\rho_j v^D + j_j = -D\frac{\partial \rho_j}{\partial r}$ and $c_j v^{D,N} + J_j^N = -D\frac{\partial c_j}{\partial r}$ the boundary conditions become

$$- \left[\rho_j v^{\Gamma,evap} - D\frac{\partial \rho_j}{\partial r} \right]_{r=R} = \Omega_j \tag{3.92}$$

$$- \left[c_j v^{\Gamma,evap} - D\frac{\partial c_j}{\partial r} \right]_{r=R} = \Omega_j^N. \tag{3.93}$$

3.3.3 Spray Polymerisation - QSSA

Spray polymerisation differs from other reactive drying processes in that the average molar weight of the polymer is not constant. The general continuum equations remain unchanged. Diffusion, however, needs further consideration and diffusive fluxes necessarily have to be calculated in a molar frame of reference. A calculation of the molar mass of the polymer is mandatory in order to obtain the correct transport behaviour. Applying the quasi-steady-state assumption, hence, the polymer's partial density needs to be calculated additionally to the concentrations of all components in order to obtain its molar weight. With the reaction terms being the same as in the lumped model (section 3.2.2) the following set of equations can be obtained:

$$\frac{\partial [S]}{\partial t} = -\frac{1}{r^2}\frac{\partial}{\partial r}\left(r^2 [S] v^N + r^2 J_S^N\right) \tag{3.94}$$

$$\frac{\partial [I]}{\partial t} = -\frac{1}{r^2}\frac{\partial}{\partial r}\left(r^2 [I] v^N + r^2 J_I^N\right) - k_d [I] \tag{3.95}$$

$$\frac{\partial [M]}{\partial t} = -\frac{1}{r^2}\frac{\partial}{\partial r}\left(r^2 [M] v^N + r^2 J_M^N\right) - k_p [M] [R_{tot}] \tag{3.96}$$

$$\frac{\partial [P]}{\partial t} = -\frac{1}{r^2}\frac{\partial}{\partial r}\left(r^2 [P] v^N + r^2 J_P^N\right) + (k_{td} + 0.5 k_{tc}) [R_{tot}]^2$$
$$+ (k_{trm} [M] + k_{trs} [S]) [R_{tot}] \tag{3.97}$$

$$\frac{\partial \rho_P}{\partial t} = -\frac{1}{r^2}\frac{\partial}{\partial r}\left(r^2 \rho_P v^N + r^2 \frac{\rho_P}{[P]} J_P^N\right) + r_P^F MW_P^{inst}$$
$$= -\frac{1}{r^2}\frac{\partial}{\partial r}\left(r^2 \rho_P v^N + r^2 \frac{\rho_P}{[P]} J_P^N\right) + k_p [M] [R_{tot}] MW_M. \tag{3.98}$$

The zero-gradient boundary condition holds for the droplet centre. At the interface to the drying gas, equation 3.90 is the boundary conditions for all concentrations. Considering the partial density, convection and diffusion are expressed with respect to the molar averaged velocity and the molar diffusive flux of the polymer so that instead of the mass based boundary condition 3.89 a reformulation of equation 3.90 has to be used

$$\left[\rho_P\left(v^{D,N} - v^{\Gamma,evap}\right) + \frac{\rho_P}{[P]} J_P^N\right]_{r=R} = 0. \tag{3.99}$$

This equation (with no evaporation of the polymer) is the same condition as for the polymer concentration

$$\left[[P]\left(v^{D,N} - v^{\Gamma,evap}\right) + J_P^N\right]_{r=R} = 0. \tag{3.100}$$

When boundary conditions are implemented as algebraic constraints, this cannot be done on both polymer conditions simultaneously, as the system of equations would be underdetermined. Either the concentration or the partial density of the polymer at the boundary therefore necessarily has to be calculated using a differential equation.

In case of (pseudo-)binary Fickian diffusion, the set of equations can be simplified, again. Using equations 3.23 or 3.30 and 3.31 the following relations are obtained

$$\frac{\partial c_j}{\partial t} = -\frac{1}{r^2}\frac{\partial}{\partial r}\left(r^2 c_j\left(\vec{v}^R - D\frac{[P]}{\rho_P^0}\frac{\partial}{\partial r}\frac{\rho_P}{[P]}\right) - r^2 D\frac{\partial c_j}{\partial r}\right) + r_j^F \tag{3.101}$$

$$\frac{\partial \rho_P}{\partial t} = -\frac{1}{r^2}\frac{\partial}{\partial r}\left(r^2 \rho_P\left(\vec{v}^R - D\frac{[P]}{\rho_P^0}\frac{\partial}{\partial r}\frac{\rho_P}{[P]}\right) - r^2\frac{\rho_P}{[P]}D\frac{\partial [P]}{\partial r}\right)$$
$$+ k_p [M][R_{tot}]MW_M. \tag{3.102}$$

The boundary conditions to the drying gas are then

$$\left[c_j\left(-D\frac{[P]}{\rho_P^0}\frac{\partial}{\partial r}\frac{\rho_P}{[P]} - v^{\Gamma,evap}\right) - D\frac{\partial c_j}{\partial r}\right]_{r=R} = \Omega_j^N \tag{3.103}$$

$$\left[\rho_P\left(-D\frac{[P]}{\rho_P^0}\frac{\partial}{\partial r}\frac{\rho_P}{[P]} - v^{\Gamma,evap}\right) - \frac{\rho_P}{[P]}D\frac{\partial [P]}{\partial r}\right]_{r=R} = 0. \tag{3.104}$$

Reaction Driven Transport in Polymerisation

The reaction induced velocity contribution is related to the molar weights of the components currently generated/consumed (equation 3.21). As has been shown in section 3.2.2, due to stoichiometry the reaction related volume change is independent from the chain length and only corresponds with the formation rate of monomer or, in good approximation, the rate of the propagation reaction. As all other reactions have a negligibly small effect on the volume, the reaction driven velocity v^R can be calculated according to

$$\frac{\partial v^R}{\partial r} = \sum r_j^F \frac{MW_j^{inst}}{\rho_j^0} = r_M^F MW_M \left(\frac{1}{\rho_M^0} - \frac{1}{\rho_P^0}\right). \tag{3.105}$$

The zero-velocity condition 3.79 is the boundary condition for solving this one-dimensional ordinary differential equation / boundary value problem.

3.3.4 Spray Polymerisation - Method of Moments

With the diffusive fluxes of the moments defined as in equation 3.44, the moments' transport equation inside the droplet become

$$\frac{\partial \zeta_k}{\partial t} = -\frac{1}{r^2}\frac{\partial}{\partial r}\left(r^2 \zeta_k v^N + r^2 \frac{\zeta_k}{\lambda_0 + \zeta_0}J_P^N\right) + r_{\zeta_k}^F \qquad (3.106)$$

$$\frac{\partial \lambda_k}{\partial t} = -\frac{1}{r^2}\frac{\partial}{\partial r}\left(r^2 \lambda_k v^N + r^2 \frac{\lambda_k}{\lambda_0 + \zeta_0}J_P^N\right) + r_{\lambda_k}^F. \qquad (3.107)$$

Like in the lumped model, the equations of other species are the same as in the QSSA, with $[R_{tot}]$ being replaced by the zeroth moment of the living chains' distribution λ_k. At the droplet centre again zero-gradients are assumed. The interface condition to the drying gas is equation 3.90, which is for the moments

$$\left[\zeta_k\left(v^{D,N} - v^{\Gamma,evap}\right) + \frac{\zeta_k}{\zeta_0 + \lambda_0}J_P^N\right]_{r=R} = 0 \qquad (3.108)$$

$$\left[\lambda_k\left(v^{D,N} - v^{\Gamma,evap}\right) + \frac{\lambda_k}{\zeta_0 + \lambda_0}J_P^N\right]_{r=R} = 0. \qquad (3.109)$$

Again these boundary conditions are linearly dependent. A formulation for bulk moments is straightforward by replacing $\lambda_k + \zeta_k$ and ζ_k with the respective bulk moment μ_k. The reaction induced velocity can be evaluated in the same way as for the QSSA employing equation 3.105, as all monomer molecules consumed by reactions are incorporated into polymer chains and $r_{\zeta_1}^F + r_{\lambda_1}^F = -r_M^F$.

The special case of (pseudo-)binary Fickian diffusion leads to

$$\frac{\partial c_j}{\partial t} = -\frac{1}{r^2}\frac{\partial}{\partial r}\left(r^2 c_j\left(v^R \bar{v}^{MW}\right) - r^2 D\frac{\partial c_j}{\partial r}\right) + r_j^F, \qquad (3.110)$$

$$\frac{\partial *}{\partial t} = -\frac{1}{r^2}\frac{\partial}{\partial r}\left(r^2 *\left(v^R \bar{v}^{MW}\right) - r^2 \frac{*}{\lambda_0 + \zeta_0}D\frac{\partial}{\partial r}\left(\lambda_0 + \zeta_0\right)\right) + r_*^F \qquad (3.111)$$

$$* = \zeta_k \text{ or } \lambda_k, \qquad \bar{v}^{MW} = -MW_M \frac{\lambda_0 + \zeta_0}{\rho_P^0}D\frac{\partial}{\partial r}\frac{\lambda_1 + \zeta_1}{\lambda_0 + \zeta_0}$$

with the boundary conditions

$$\left[c_j\left(\bar{v}^{MW} - v^{\Gamma,evap}\right) - D\frac{\partial c_j}{\partial r}\right]_{r=R} = \Omega_j^N \qquad (3.112)$$

$$\left[*\left(\bar{v}^{MW} - v^{\Gamma,evap}\right) - \frac{*}{\lambda_0 + \zeta_0}D\frac{\partial}{\partial r}\left(\lambda_0 + \zeta_0\right)\right]_{r=R} = 0 \qquad (3.113)$$

$$(3.114)$$

3.4 Comparison with Existing Models

The concept of diffusion and reaction driven convection is not completely common in drying models. Similar formulations can be found at Czaputa and Brenn (2012), Handscomb, Kraft, and Bayly (2009), and Seydel (2005). Other contributions, such as (Brenn 2004; Sloth et al. 2006), omit the diffusive velocity contribution. As very many drying models only involve two phases, the problem is simplified and can be solved by elimination of the second phase and insertion (like in Czaputa and Brenn 2012). Seydel calculates transport of a solid phase within a solvent by means of population balances. By summation of the volume fluxes of both phases up to zero, he expresses the corrective velocity depending on the mass fractions (practically the same expression can be found in Handscomb, Kraft, and Bayly 2009):

$$v_r = D \frac{\partial w_G}{\partial r} \frac{\frac{1}{\rho_G^0} - \frac{1}{\rho_L^0}}{\frac{w_G}{\rho_G^0} + \frac{w_L}{\rho_L^0}}. \tag{3.115}$$

This approach can be extended to more components and rewritten for Fickian diffusion in a molar notation using mole fractions and molar weights.

In comparison, the corrective convection contributions in this work are directly derived from continuum laws and condensed in short, elegant equations. The approach is general in terms of the number of components and molar or mass based notation. The effect of reaction driven density changes is included. Moreover, Maxwell-Stefan equations can be implemented in a straightforward manner. Finally, it allows for a varying (averaged) molar weight, which admittedly is a special case, but important when diffusion of polymers with locally changing degrees of polymerisation is to be modelled.

Generally speaking, drying models always need to consider a convective contribution, even if only diffusive transport is present, as mass conservation is violated otherwise. A formulation with respect to partial densities is to be preferred. The equation of the diffusive velocity becomes somewhat cumbersome regarding to mass fractions, especially for a higher number of components, but is simple and straightforward concerning partial densities. If only (pseudo-)binary Fickian diffusion is involved, the latter reduces to simply Fick's second law (see equations 3.11 and 3.13). This can be applied to the great majority of drying models and provides mass conservation and a very simple formulation at the same time.

3.5 Implementational Considerations

The system of partial differential equations can be numerically solved using the method of lines. The radial coordinate is discretised by an appropriate method - like finite differences (FDM) or finite volume methods (FVM). In doing so, an initial value problem (IVP) is obtained, which consists of either solely ordinary differential equations or a mixture of differential equations in the fluid bulk and algebraic equations at the boundary. This system of equations can be solved by standard ODE or DAE solvers. Due to the different time scales of physical and chemical processes, the problem is stiff and cannot be treated efficiently by every class of methods for IVP solving. BDF (backward differentiation formula) solvers are suited well and commonly used for stiff problems.

The droplet polymerisation model equations exhibit some peculiarities, by which the implementation is non-trivial. These obstacles will be addressed in the following. There is ample literature on the above-mentioned methods. Comparing finite differences and finite volume methods, higher order discretisations can easily be derived in the FDM, which is of advantage when strong gradients occur near the interface to the drying gas. Moreover, the moving boundary system of the shrinking droplet just involves one simple, additional term. On the other hand, the FVM is advantageous with respect to conservation and the implementation of the Neumann boundary conditions at the drying gas is straightforward. Numerical solutions in this work have been obtained by a Python implementation of the model using a finite volume discretisation and the standard BDF/NDF solver of Python's SciPy library (Virtanen et al. 2020).

3.5.1 Implementation of the Moving Boundary Problem

The droplet is shrinking steadily due to evaporation and may change its size as well because of chemical reactions. Therefore, the motion of the interface to the drying gase needs to be considered in the implementation (equations 3.88 and 3.91). This can either be implemented in ways that discretisation points are fixed and the interface moves over these points or by an adaptive computational domain (Crank 1987, p. 163). The first approach is more complicated with respect to the implementation of the boundary conditions, as the points being involved into the boundary conditions as well as their discretisation stamp are continually changing, and prone to numerical errors. Moreover, the solution via an adaptive grid is not only more elegant, but easily implemented for one-dimensional problems like droplet drying. This can be undertaken by a coordinate transformation of the radial coordiante r to a dimensionless coordinate ξ, which spans between the droplet centre and the droplet boundary:

$$y(r,t) \longrightarrow y(\xi,t) \qquad \text{with} \quad \xi(r=0) = 0, \; \xi(r = R(t)) = 1. \qquad (3.116)$$

$$\xi = \frac{r}{R(t)} \qquad (3.117)$$

$$\frac{\partial \xi}{\partial r} = \frac{1}{R(t)} \qquad (3.118)$$

$$\frac{\partial \xi}{\partial t} = -\frac{r}{R(t)^2} \frac{dR}{dt} = -\frac{\xi}{R(t)} \frac{dR}{dt} \qquad (3.119)$$

Thereby, the following conversion formulae for the calculation of a quantity y can be obtained (cmp. Crank 1987, p. 170):

$$\frac{\partial y}{\partial r} = \frac{\partial y}{\partial \xi} \frac{\partial \xi}{\partial r} = \frac{\partial y}{\partial \xi} \frac{1}{R(T)} \qquad (3.120)$$

$$\left.\frac{\partial y}{\partial t}\right|_r = \left.\frac{\partial y}{\partial t}\right|_\xi + \frac{\partial y}{\partial \xi} \frac{\partial \xi}{\partial t}. \qquad (3.121)$$

The temporal derivative of this quantity considering the moving coordinate ξ is

$$\left.\frac{\partial y}{\partial t}\right|_\xi = \left.\frac{\partial y}{\partial t}\right|_r + \frac{\partial y}{\partial \xi} \cdot \frac{\xi}{R(t)} \frac{dR}{dt}. \qquad (3.122)$$

The first term on the right hand side denotes the Eulerian continuum law, the transport equation of the respective quantity. The second term accounts for the motion of the coordinate ξ. Formally, the course of action is similar to applying a Lagrangian frame of reference. However, observer's velocity is not determined by the fluid but by the motion of the coordinate with the velocity $\xi \frac{dR}{dt}$.

3.5.2 Boundary Conditions

The boundary conditions 3.90, 3.108 and 3.108 are algebraic equations. If the gradients are discretised by finite differences, these equations can be calculated simultaneously to the differential equations in the fluid bulk by a DAE solver. Whereas this works perfectly well for normal components using equation 3.90, the system becomes singular when higher moments are considered by equations 3.108 and 3.109. As has been stated before, the moments' boundary conditions are linearly dependent. Factorising the moments ζ_k and λ_k , only one single polymer boundary condition remains:

$$\left[v^{D,N} - v^{\Gamma,evap} + \frac{J_P^N}{\zeta_0 + \lambda_0} \right]_{r=R} = 0. \tag{3.123}$$

It just follows from the ansatz of all moments behaving the same way with respect to transport - as one single component - that the moments' balance at the boundary is condensed into one equation. In case of finite differences polymer boundary conditions therefore cannot be implemented as algebraic equations, but need to be incorporated into the discretised partial differential equations at the interface nodes. This is achieved by replacing $v^{D,N} + \frac{J_P^N}{\zeta_0 + \lambda_0}$ at position $r = R$ by the interface velocity $v^{\Gamma,evap}$ in the discretisation of the divergence term. For other components $c_j v^N + J_j^N$ can be replaced with $c_j v^{\Gamma} + \Omega_j^N$ so that solely differential equations need to be solved at the outer boundary.

A finite volume implementation of these boundary conditions is in contrast straightforward. The domain's boundary is identical to the outmost cell interface. Fluxes over this cell boundary are hence defined by the fluxes over the interface to the drying gas obtained by linear driving forces. In case of the non-volatile polymer, there is just a zero flux accross the outmost cell interface.

3.5.3   Treatment of Convection Terms

Despite the absence of directed fluid motion, a convective flux arises inside the droplet due to the different specific volumes of the diffusing species and as a result of density changing reactions (equations 3.20 and 3.21). Whenever convective terms are discretised by higher order schemes, the computation is prone to numerical oscillations. In a second order finite difference scheme the gradient at a point i is calculated using values of the neighbouring points $i-1$ and $i+1$ so that the gradient of a sawtooth like profile is calculated to zero everywhere. The same holds for finite volume approximations, when interface values between two cells are calculated by the arithmetic mean of both cell values. Such oscillations typically origin at a discontinuity (Hirsch 1990, p. 408) like the interface to the drying gas and - not being damped - may grow throughout a computation until spatial profiles become distorted.

This can be overcome by upwinding using first order approximations, i.e. obtaining information from upstream points. A finite difference gradient then involves values of points $i-1$ and i if the velocity points in positive direction. Likewise, interface values in a finite volume calculation are taken from the cell opposite to the flow direction - the donor cell approach. Such first order approximations involve for their part numerical diffusion. Sharp gradients are artificially flattened out depending on the numerical grid size.

Upwinding schemes are not symmetric and depend on the flow direction. The sign of the diffusion induced convective flux may change throughout a computation. One example is initial droplet expansion by cause of condensation at high saturation of the drying gas followed by shrinkage due to drying after droplet heat up. Therefore, an upwinding discretisation in droplet polymerisation needs to adapt to locally and temporally varying flow directions.

The disadvantage of numerical diffusion can be countered by monotonicity preserving discretisations of higher order, so called TVD-schemes (Total Variation Diminishing). This approach has been defined by Harten (1983), whereas the basic concept has already been laid out by van Leer (1973). Essentially, a limiter function switches between first and second (or higher) order approximations, depending on the local degree of numerical oscillations. Numerical diffusion is prevented as far as possible and oscillations are effectively damped.

A trivial approach for countering numerical diffusion is to use a larger number of grid points. As this is costly, it is advisable to provide a tight discretisation only in regions of steep gradients. In droplet drying this affects the drop's outer rim near the interface to the drying gas. A simple rule for a refined discretisation is to set grid points or volumes not equispaced but such that the volume represented by each node is equal.

These numerical challenges are no special characteristic of spray polymerisation, but will naturally arise in diffusion dominated systems, when specific volumes are very different and diffusion induced convection is to be considered. In a mass based notation the differences of bulk densities are however typically not that large that the convective term becomes dominant and numerical fluctuations cannot be damped by diffusion. As a molar based solution additionally involves the molecular weight, specific volumes can differ to a much larger degree, especially in case of macromolecules. The countermeasures - first order upwinding or TVD schemes - are common knowledge though. Still, diffusive moment transport according to equation 3.43 introduces additional peculiarities.

Diffusion of higher moments is not calculated by a typical diffusion term, but is in fact convection with the diffusion contribution $\frac{\tilde{J}_P^N}{\zeta_0 + \lambda_0}$ of the zeroth moment's species velocity. If this value exceeds the fluid's convection and is oppositely directed, the upwinding discretisation of the ordinary convective term will actually consider downstream values when applied to higher moments' diffusion. As a result, the motion of higher moments is calculated incorrectly and the computation becomes unstable. The first order/upwinding discretisation of the higher moments' diffusion has therefore to be set seperately from the ordinary convective term based on the direction of the moments' diffusive flux.

When diffusion is modelled in a pseudo-binary Fickian way, the use of an extended Fick's law (3.23) can alleviate the problem of numerical oscillations. The molar weight induced velocity contribution is typically significantly smaller than the ordinary diffusion driven convective contribution which vanishes in Fick's second law. The convective part of the transport equation is hence reduced so that the overall behaviour becomes largely dominated by diffusion.

Appropriate Application of TVD Limiters

Numerical tests concerning a TVD scheme employed the van Leer limiter $\phi^{vanLeer}$ (van Leer 1973) in this work. A value $u_{i+1/2}$ between two finite volume cells i and $i+1$ is calculated as follows:

$$u_{i+1/2}^{TVD} = \left(1 - \phi_{i+1/2}^{vanLeer}\right) u_{i+1/2}^{1st} + \phi_{i+1/2}^{vanLeer} u_{i+1/2}^{2nd} \qquad (3.124)$$

$$\phi_{i+1/2}^{vanLeer} = \frac{\theta_{i+1/2} + |\theta_{i+1/2}|}{1 + |\theta_{i+1/2}|} \qquad (3.125)$$

$$\theta_{i+1/2} = \frac{u_i - u_{i-1}}{u_{i+1} - u_i} \cdot (v_{i+1/2} \geq 0) + \frac{u_{i+1} - u_i}{u_{i+2} - u_{i+1}} \cdot (v_{i+1/2} < 0). \qquad (3.126)$$

The limiter function depends on the local degree of oscillation θ. Whether $u_{i+1/2}^{1st}$ equals u_i or u_{i+1} depends on the velocity direction. The second order approximation $u_{i+1/2}^{2nd}$ corresponds with an arithmetic mean (weighted in case of uneven point distribution).

Limiting has to act the same way for all components. If inter-cell values in an FVM implementation of various components are computed via different orders, conservation will be violated. In each cell, the sum of all volume fractions is one. Applying the donor cell approach to all components involves only one cell's values and is therefore consistent. Also a second order approximation -

the arithmetic mean of two cells' values - will again sum up to one. If however, the inter-cell value of one component is obtained from the donor cell and values of other components via arithmetic mean, the sum of all volume fractions will commonly be different to one and convection wil violate conservation. The limiter function hence needs to evaluate the largest oscillations of all species and apply the lowest order to all components.

3.5.4 Implementation of Diffusion

Diffusion involves the calculation of inter-cell values. In the simple case of binary Fickian diffusion

$$\vec{J}_j^N = -Dc\nabla x_j = -D\nabla c_j + Dx_j\nabla \ln c$$

either the overall concentration c or the mole fraction x_j need to be known at the boundary between two finite volume cells (or at a staggered grid point in an FDM discretisation). Approximation via arithmetic mean can be used in this case, as the concentration is not multiplied with a fixed velocity but individual mole fractions' gradients. These gradients will act contrary to oscillations of a species' concentration as long as all concentrations are consistent, i.e. all volume fractions add up to one.

Higher moments' diffusion is different, as - from a numerical point of view - their motion is purely convective and does not involve a truely diffusive term. Only the polymer's / zeroth moments' diffusive flux $\vec{J}_P^N$ is known at inter-cell positions, while other concentrations in higher moments diffusion

$$\vec{J}_{\zeta_j}^N = \frac{\zeta_j}{\zeta_0 + \lambda_0}\vec{J}_P^N$$

need to be approximated. Interpolation of inter-cell values via arithemtic mean hence acts like a second order convective term with respect to the velocity $\frac{\vec{J}_P^N}{\zeta_0 + \lambda_0}$ and can be prone to numerical oscillations. If on the other hand upwinding is applied, inter-cell concentrations via arithmetic mean and donor cell values are mixed, which violates conservation. This violation is however typically small. If necessary, an additional volume correction can be implemented based on the constraints of the sum of volume fractions being one or its change over time being zero. The deviation from either condition can be used for a corrective velocity, calculated similarly to the reaction induced velocity as a boundary value problem (see appendix B for details).

3.6 Verification of the Transport Approach

The concept of diffusion and reaction driven convection as derived above will be applied to various test cases and verified in simple numerical experiments in the following, before simulations of the overall (reactive) drying process will be presented in the next chapter. All test cases are based on a one-dimensional description, which corresponds with the dimensionality of the drying models. An extension to higher dimensions is straightforward for the diffusion induced velocity contribution.

3.6.1 Diffusion Driven Convection, Constant Properties

A very simple test case is binary Fickian diffusion of two components A and B. Both components differ in density and molar weight and, accordingly, in their mass specific volume v^m and molar volume v^N. The domain consists of a volume stretching from $z = 0\,\text{m}$ to $z = L = 1\,\text{m}$. The boundaries are assumed to be impermeable, which corresponds with Neumann conditions of $\nabla w_j\big|_{z=0,z=1} = \nabla \rho_j\big|_{z=0,z=1} = \nabla c_j\big|_{z=0,z=1} = 0$. Initially, the left part of the volume is filled with pure component B and the right one with pure A. The diffusion coefficient is chosen to $D = 1 \times 10^{-9}\,\text{m/s}^2$ (which however only affects the time-scale of the dynamic problem). Each graph provides simulated profiles between 0 and $1 \times 10^9\,\text{s}$, with a stepping of $2.5 \times 10^7\,\text{s}$.

Figure 3.2 shows profiles, when writing the transport equation in terms of the mass fraction w_A and neglecting the diffusion related velocity contribution

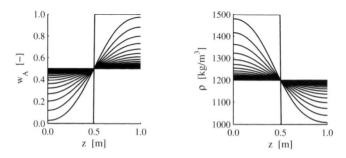

Figure 3.2: Binary diffusion, transport equation in w_A, diffusion induced convection neglected.

Table 3.1: Test case binary diffusion, physical data.

	A	B
density $\left[\frac{kg}{m^3}\right]$	1000	1500
molar weight $\left[[10^{-3}\frac{kg}{mol}\right]$	18	180
$v^m \left[10^{-3}\frac{m^3}{kg}\right]$	1	0.667
$v^N \left[10^{-3}\frac{m^3}{mol}\right]$	0.018	0.120

(as a drying model compare Sloth et al. 2006):

$$\frac{\partial w_A}{\partial t} = \nabla\left(D\nabla w_A\right).$$

Under the considered initial conditions, a system will always relax to $w_A = w_B = 0.5$ as in Figure 3.2, if the convective part is not regarded. With the pure density of B being 1.5 times greater than the one of A the final mass fraction of component A should be 40 %, when two equal volumes of A and B are mixed, and the final mixture density 1250 kg/m^3.

This also holds, if not mass or mole fractions are balanced but partial densities or concentrations, as in Figure 3.3 with the transport equation

$$\frac{\partial c_A}{\partial t} = -\nabla \vec{J}_A^N, \quad \vec{J}_A^N = -cD\nabla x_A.$$

The mole fraction of component A is equilibrated to its correct value (~ 0.87), but the concentrations are not calculated correctly.

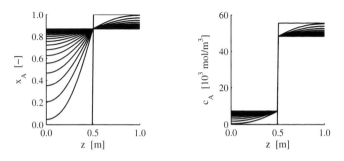

Figure 3.3: Binary diffusion, transport equation in c_A, diffusion induced convection neglected.

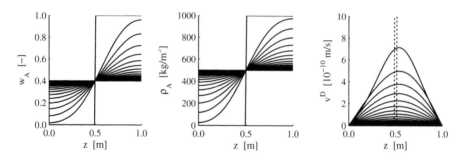

Figure 3.4: Binary diffusion, transport equation in w_A or ρ_A, diffusion induced convection considered.

Addition of the convective part (see also equation 3.9) leads to the equations

$$\frac{\partial w_A}{\partial t} = 2D\frac{\nabla \rho}{\rho}\nabla w_A + \nabla\left(D\nabla w_A\right)$$

$$\frac{\partial c_A}{\partial t} = -\nabla\left(c_A\vec{v}^{D,N} + \vec{J}_A^N\right), \quad \vec{v}^{D,N} = -\sum \frac{\vec{J}_j^N MW_j}{\rho_j^0}$$

and solves the problem correctly, as shown in Figures 3.4 and 3.5. The same results can be obtained using Fick's second law $\frac{\partial c_A}{\partial t} = \nabla\left(D\nabla c_A\right)$. Due to the stepwise initial species distribution, initial velocity profiles (dashed lines) exhibit a Dirac delta function like peak at the transition between the volumes containing solely A and B. The diffusion induced velocity is indeed dependent on the frame of reference. With the mass specific volume v_A^m of component A being higher

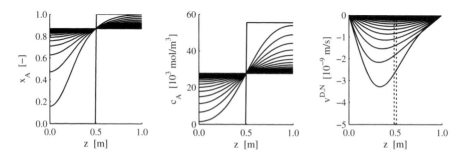

Figure 3.5: Binary diffusion, transport equation in c_A, diffusion induced convection considered.

than the one of species B, but the molar specific volume v_A^N being smaller, the velocity profiles of the molar and mass averaged solutions are of different signs and values. The more the specific volumes of mixture components differ, the higher the magnitude of the corrective velocity is. This becomes especially important when modelling diffusion of macromolecules in molar notation and thus markedly different molar volumes of the various species.

3.6.2 Diffusion Driven Convection, Variable Molar Weight

Considering typical applications in which diffusive transport is modelled, like drying, a species having variable (averaged) molar weight is rather exotic. A molar mass transport model will primarily be used for reactive processes involving polymers, in which the time and position dependence of the polymer properties is of interest. Nevertheless, as a validation example a polymer membrane consisting of two layers I and II may be considered, which is infiltrated by solvent (S). If both layers are made up of the same polymer (P), but with a different degree of polymerisation/chain length, this case can be modelled as binary diffusion with the polymer species having a variable molar weight. The species' values are provided in Table 3.2. As a sharp jump is hard to be resolved numerically, the polymer's molar weight changes continuously near the transition between the layers I and II. The initial polymer concentration is equal throughout the complete domain with a value of 90 % of pure polymer with the larger molar weight and hence smaller bulk concentration $c_P^{0,I} = \frac{\rho_P^0}{MW_P^I}$. Due to its smaller molar weight in layer II, the polymer's bulk concentration is twice as high there compared to layer I. Hence, a larger amount of water is initially contained in compartment I compared to layer II. The initial slope of c_S can be determined by the closing condition $c_S = \frac{1}{v_S^N}\left(1 - c_P v_P^N\right) = \frac{\rho_S^0}{MW_S}\left(1 - c_P\frac{MW_P}{\rho_P^0}\right)$. Figure 3.6 shows the initial profiles as dashed lines. Again, the domain is bounded at both sides. The solvent contained for a larger part in layer II therefore penetrates into layer I until equilibrium is achieved. All numerical results have been obtained by solving the transport equations for c_S, c_P and ρ_P. This is equivalent to a solution with respect to polymer moments as $\zeta_1 MW_M = \rho_P$.

The naive assumption of diffusive polymer transport by Fick's law

$$\frac{\partial c_j}{\partial t} = \nabla\left(D\nabla c_j\right) \qquad \frac{\partial \zeta_s}{\partial t} = \nabla\left(D\nabla\zeta_s\right) \qquad \frac{\partial \rho_j}{\partial t} = \nabla\left(D\nabla\rho_j\right)$$

Table 3.2: Test case binary diffusion with variable molar weight, physical data.

	S	P^I	P^{II}
density $\left[\frac{kg}{m^3}\right]$	1000	1200	1200
molar weight $\left[10^{-3}\frac{kg}{mol}\right]$	18	400000	200000
$v^m \left[10^{-3}\frac{m^3}{kg}\right]$	1	0.833	0.833
$v^N \left[10^{-3}\frac{m^3}{mol}\right]$	0.018	333.3	166.7

involves diffusion of higher moments or the polymer's partial density (cmp. sections 3.1.4 and 3.1.5). This means nothing else than that not only the mole fractions are equilibrated, but also the polymer's molar mass will be smoothed out. The profiles in Figure 3.6 confirm this prediction. The initial jump of the molar weight is completely equilibrated at the end of the calculation. In the con-

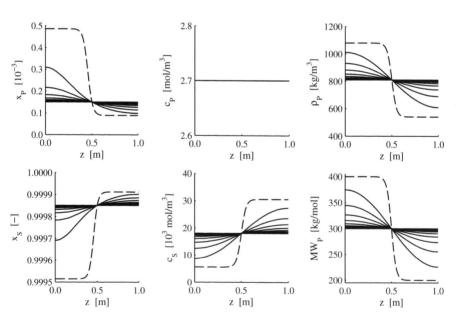

Figure 3.6: Binary diffusion with variable molar weight, transport equation using Fick's second law, profiles of the components and the polymer's molar mass with dashed lines showing the initial state.

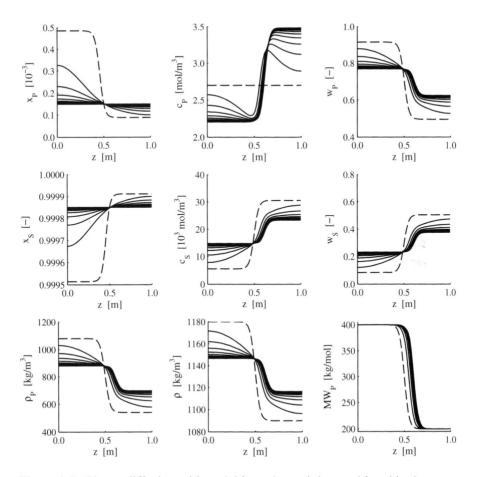

Figure 3.7: Binary diffusion with variable molar weight considered in the transport equation, dashed lines show the initial state.

sidered application, both membrane layers would hence diffuse into each other, resulting in a mixed polymer of constant average molecular weight.

The corrected transport equations according to Fick's second law for variable molar weights are (see equations 3.22, 3.23 and 3.25)

$$\frac{\partial c_j}{\partial t} = \nabla \left(c_j D \frac{c_P \nabla MW_P}{\rho_P^0} + D \nabla c_j \right)$$

$$\frac{\partial \rho_P}{\partial t} = \nabla \left(\rho_P D \left(\frac{c_P \nabla MW_P}{\rho_P^0} + \frac{\nabla c_P}{c_P} \right) \right).$$

The unphysical smoothing of the polymer's molar weight is avoided by coupling the transport of its partial density to the molar frame of reference. Additionally, the additional convective term corrects Fick's second law, which is only valid for species having constant physical properties. The same result could have been obtained by to the general expressions

$$\frac{\partial c_j}{\partial t} = -\nabla\left(c_j \vec{v}^{D,N} + \vec{J}_j^N\right) \qquad \frac{\partial \rho_P}{\partial t} = -\nabla\left(\rho_P \vec{v}^{D,N} + MW_P \vec{J}_j^N\right)$$

with $\vec{J}_i^N = -cD\nabla x_i$. The convective contribution due to molar mass differences is included in the diffusion induced velocity $\vec{v}^{D,N} = -\sum \frac{\vec{J}_j^N MW_j}{\rho_j^0}$ and has not to be considered separately (see section 3.1.3). As can be seen from Figure 3.7, the molar mass of the polymer is preserved at its initial values. Unphysical polymer mixing is avoided. The transition between both layers moves towards the right, which is a consequence of solvent transport from layer *II* to layer *I*. The polymer within the left section is swelling, wheras the polymer in layer *II* is shrinking. Unlike in other binary diffusion systems, only the mole fractions are equilibrated. Due to the molar mass differences, the initially constant polymer concentration is steepening, as the solvent concentration raises in layer *I* and shrinks in the second section. The solvent concentration is only partially equilibrated. A very important observation is that even the mass fractions are not constant. A formulation of Fickian diffusion in a mass based frame of reference would result in wrong profiles, unless fluxes and diffusion induced velocities were calculated according to equations 3.26 and 3.27.

An extension of this example to higher moments is straightforward.

3.6.3 Diffusion Driven Convection, Excess Volumes

Even if a varying molar mass is considered in the previous example, the solution is ideal in that mixing does not induce any volume effect. However, excess volumes may occur in real solutions, which means that the specific reference volumes of the components depend on the (local) mixture composition. This effect shall in principle be studied by a simple example. The basic initial setup is the same as in the first test case, with two components A and B being initially separated. The physical properties of the pure species are identical as well. However, the density of component *A* is not taken constant, but as a function of the mass fraction of B

$$\rho_A^0 = \rho_A^{0*} + w_B \frac{d\rho_A^0}{dw_B}.$$

In order to provide clear graphical results, a value of $\frac{d\rho_A^0}{dw_B} = 400\,\text{kg/m}^3$ is considered, which is very large in comparison to the pure reference density of $\rho_A^{0*} = 1000\,\text{kg/m}^3$. As equal volumes of species A and B are initially provided, the mass fractions after equilibration are $w_A^{eq} = 0.4$ and $w_B^{eq} = 0.6$. The final reference density of component A is therefore $1240\,\text{kg/m}^3$ so that the volume occupied by A contracts by a ratio of $\frac{1000}{1240} = 0.8065$. As half of the initial setup consists of pure component A, the overall domain shrinks by a ratio of 0.9032.

As has been pointed out in section 3.1.2, obtaining direct, analytical expressions for diffusion and reaction induced velocities becomes difficult under the occurence of excess volumes. Here, the derivative of the closing condition 3.14 is

$$\sum_j \frac{\partial\left(\rho_j/\rho_j^0\right)}{\partial t} = \sum_j \frac{1}{\rho_j^0}\frac{\partial\rho_j}{\partial t} - \frac{w_A}{\rho_A^{0^2}}\frac{d\rho_A^0}{dw_B}\left(\frac{\partial\rho_B}{\partial t} - w_B\frac{\partial\rho}{\partial t}\right) = 0.$$

This equation can be applied as an algebraic constraint for the calculation of the volume corrective velocities. If the left boundary is considered as being fixed, its velocity is zero and the boundary conditions for components A and B are identical to those in the ideal case (section 3.6.1). The right boundary at $z = L$ is moving due to the contraction of the domain. By splitting the total velocity v into the standard diffusion induced part $v^D = -\sum \frac{j_j}{\rho_j^0}$ and the unknown part v^P resulting from the varying reference density of A, it can be easily shown that the motion of the boundary v^Γ equals v^P. Division of the general boundary condition (the same as formula 3.80 for non-volatile components in droplet drying)

$$\rho_j\left(v - v^\Gamma\right) + j_j = 0$$

by ρ_j^0 and summation over all components leads to

$$\sum_j \frac{\rho_j}{\rho_j^0}\left(v^D + v^P - v^\Gamma\right) + \sum_j \frac{j_j}{\rho_j^0} = \left(v^D + v^P - v^\Gamma\right) - v^D = 0$$

$$\longrightarrow \quad v^\Gamma = v^P.$$

The boundary condition for a single component is $\rho_j v^D + j_j\big|_{z=L} = 0$, which is the same as for a fixed, impermeable wall and simplifies for binary Fickian diffusion to $\nabla \rho_j|_{z=L} = 0$. In the numerical solution of the problem, the moving boundary problem was implemented by a coordinate transformation as explained in section 3.5.1. The algebraic velocity equation was only solved for v^ρ, whereas the diffusion induced velocity was explicitly treated. As can be seen from Figure 3.8, the mass fraction is equilibrated to 0.4. However, in comparison to the initial example with constant reference densities (section 3.6.1), the partial density evolves to higher values due to the volume effects in the mixture. Initially, the reference density ρ_A^0 exhibits a step-wise slope due to the pure species in both parts of the domain. The final reference density is in accordance with its predicted value, as is the contraction of the domain $\frac{L}{L_0}$.

The mixture density matches its theoretical value of $1250/0.9032 = 1384\,\mathrm{kg/m^3}$. Velocity profiles are provided for v^ρ and correspond with the volume contraction due to the density increase. The initial velocity profile (dotted line) is a (theoretically) infinite step function, with its magnitude in simulations depending on the numerical resolution.

In the problems considered in this work, it is generally assumed that excess

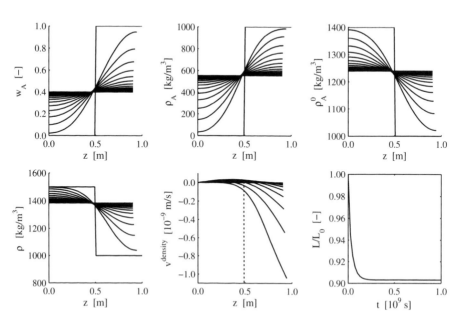

Figure 3.8: Binary diffusion considering volume effects in the mixture, transport by Fickian diffusion.

volumes are negligible and the densities of the components in a mixture are constant. Then the diffusion and reaction induced velocity contributions can be directly calculated without the detour of an additional set of algebraic equations. Still, this small example shows that problems, in which the reference densities are depending on the mixture compositions, can be consistently solved as long as there is a mathematical conjunction of the density change with mixture properties such as mass fractions.

3.6.4 Reaction Induced Convection

An ideally mixed batch reactor, in which a volume/density changing chemical reaction takes place, may serve as a test case for the approach of reaction driven flow. Typically, such problems are treated by lumped, 0D considerations due to the absence of spatial gradients. However, this test case can be solved analytically in a one-dimensional way as well.

For reasons of simplicity the reactor is considered as a cylinder with vertical walls. Mass transfer between the gas and the liquid phase is neglected. Although the detailed velocity field inside a real reactor is very complicated and density changing reactions induce in fact a three-dimensional flow field, for a reactor of constant cross-section the problem can be considered as pseudo one-dimensional, as the net effect of the velocity field of a rising or lowering fluid surface only applies in vertical direction. Due to the ideal mixing of the reactor volume the concentrations of a component are identical everywhere so that diffusion is not taking place. The component velocity is hence equal to the fluid velocity, which is just the reaction induced velocity v^R in the reduced, one-dimensional description. The component balance at the liquid surface is the same as equation 3.84, in which the interface velocity is equal to the component velocity of non-volatile species. The reaction induced convection according to equation 3.8 can, again, be treated as a boundary value problem. The whole problem is characterised as follows

$$\frac{\partial c_j}{\partial t} = -\frac{\partial \left(c_j v^R \right)}{\partial z} + r_j^F \tag{3.127}$$

$$\frac{\partial v^R}{\partial z} = \sum \frac{r_j^F MW_j}{\rho_j^0}, \qquad v^R(z=0) = 0 \tag{3.128}$$

$$\frac{dh}{dt} = v^\Gamma = v^R(z=h). \tag{3.129}$$

h is the height of the fluid level inside the reactor. With $c_j \neq f(z)$ and $r_j^F \neq f(z)$ the set of equations can be simplified by inserting equation 3.128 into 3.127 and integrating 3.128

$$\frac{\partial c_j}{\partial t} = r_j^F - c_j \sum \frac{r_j^F MW_j}{\rho_j^0} \tag{3.130}$$

$$v^R(z) = z \sum \frac{r_j^F MW_j}{\rho_j^0} \tag{3.131}$$

$$\frac{dh}{dt} = h \sum \frac{r_j^F MW_j}{\rho_j^0}. \tag{3.132}$$

A lumped, 0D description of the problem is (cmp. section 3.2.1)

$$\frac{dc_j}{dt} = r_j^F - \frac{c_j}{V}\frac{dV}{dt} = r_j^F - \frac{c_j}{h}\frac{dh}{dt} \tag{3.133}$$

$$\frac{dh}{dt} = \frac{1}{A}\frac{dV}{dt} = h \sum \frac{r_j^F MW_j}{\rho_j^0}, \tag{3.134}$$

with A and V being the reactor cross-section and volume, respectively. Both descriptions are equal, which shows that the reaction driven velocity represents the volume change throughout the reactor consistently in the spatially resolved case.

4. SIMULATION OF SPRAY POLYMERISATION

In the following, the spray polymerisation models provided in chapter 3 are analysed for the test case of acrylic acid polymerisation in droplets. Due to the limited residence time in a spray dryer, polymerisation reactions have to be fast and full conversion needs to be achieved within less than a minute. Acrylic acid reactions are of the fastest in polymerisation. Moreover, spray polymerisation of this system has already been investigated in lab scale in the work of Franke, Moritz, and Pauer (2017). It is hence a sensible choice for the theoretical investigations in this work. Kinetics have been chosen similar to the values of Wittenberg (2013). Before going into details of simulations, it needs to be clearly stated that this work is focused on modelling spray polymerisation in principle and deriving the fundamental effects of process parameters with PAA synthesis being a reasonable example system. With the current kinetic data, a detailed prediction of polymer properties is not possible. The following studies will therefore exhibit the qualitative features and interdependencies of the process rather than detailed properties of AA polymerisation.

First of all, the lumped, 0D approach for the method of moments is applied to show peculiarities of spray polymerisation. Further on, spatial effects are resolved by 1D simulations and compared to the 0D method. The influence of monomer evaporation is evaluated as well as partial pre-polymerisation before atomisation as an alternative process approach. Moreover, the influence of mixture thermodynamics and changes within the drying gas throughout the process are investigated for large monomer fractions in the drying gas. The simpler Quasi-Steady-State Assumption model is compared to the method of moments. Finally, the process variants of droplet polymerisation in a solution, in bulk and with pre-polymerisation are further investigated by means of numerical DoEs.

4.1 Kinetics and Process Conditions

Process conditions and concentrations in a spray droplet may differ significantly from kinetic experiments. Due to drying, the monomer content can be much higher in spray polymerisation, which is also a desired property as the solvent should be removed throughout the process. Finding appropriate kinetic data is therefore a challenge when spray polymerisation processes are to be simulated.

Wittenberg (2013) provides the most complete kinetic scheme on acrylic acid polymerisation in the literature, covering monomer contents up to 60 w%. Extrapolation of the data to higher monomer contents may introduce significant deviations from actual reaction kinetics in a drop. This is especially true for bulk polymerisation as the concentration dependency of propagation reactions has been derived for dilution of acrylic acid and enforces the monomer mass fraction w_M, which is rather a function of conversion than of monomer to solvent ratio at high monomer contents. Recent bulk polymerisation data has been provided by Dušička, Nikitin, and Lacík (2019), yet only for a temperature of 25 °C, with k_p being slightly above $30\,m^3/(mol\,s)$. For $w_M = 1, T = 298.15\,K$, Wittenberg's correlation (pp. 149-154) yields the following k_p value:

$$A_p = 120000 \cdot \left(0.063 + (1 - 0.063) \cdot e^{-17}\right) \approx 7560\,m^3/mol\,s$$

$$E_{a,p} = 67000 \cdot e^{-8.6} + \frac{2600}{1 + 50 \cdot e^{-9.9}} + 10400 \approx 13000\,J/mol$$

$$k_p \approx 40\,\frac{m^3}{mol\,s}.$$

k_p in bulk polymerisation is overestimated by about 30 % at low temperatures using Wittenberg's kinetics. The deviation at high temperatures remains unkown.

Large extrapolation errors with the present kinetic data is inevitable. As the aim of this work is not focused on obtaining PAA polymer properties in all details, but to obtain principle cause-effect relationships of polymerisation within a drop, the scheme was simplified as follows. Wittenberg applied several modifications to plain Arrhenius equations. In particular, the chain length dependency of termination conflicts with the basic premise of the method of moments that all chains behave the same way. Therefore, an averaged A_t was applied, which goes along with a chain length slightly below 13000 monomer units. Moreover, only ordinary moments of the chain length distribution are calculated without distinction between secondary and tertiary radicals and implementation of backbiting,

Table 4.1: Chemical reactions in PAA spray polymerisation model.

reaction mechanism		reaction equation
initiation	$2f_d M + I \longrightarrow IC + 2f_d R_1$	$k_d = A_i e^{-\frac{E_{a,I}}{\Re T}}$
propagation	$R_s + M \longrightarrow R_{s+1}$	$k_p = A_p e^{-\frac{E_{a,p}}{\Re T}}$
termination		$k_t = A_t e^{-\frac{E_{a,t}}{\Re T}}$
by recombination	$R_s + R_t \longrightarrow P_{s+t}$	$k_{tc} = (1 - f_{td}) k_t$
by disproportionation	$R_s + R_t \longrightarrow P_s + P_t$	$k_{td} = f_{td} k_t$

which is considered in Wittenberg's model with tertiary radicals' termination being about an order of magnitude slower.

The reaction scheme is summarised in Table 4.1. Kinetic parameters used in this work are provided in Table 4.2. The kinetics consist of plain second order reactions except for initiation, which is of first order. Yet, many polymerisation kinetics contain additional dependencies of termination rates on conversion (Trommsdorff-Norrish / gel effect) and of chain propagation on the initial monomer content w_M^0. Respective modifications were implemented in the model as well and could be activated by switches. Formulae are similar to the ones provided by Wittenberg (2013). As high monomer contents are typical in spray polymerisation, the k_p modification of E_a, which mostly applies to small monomer concentrations, was left out and the w_M^0 dependency of A_t was extended beyond Wittenberg's limit value of 0.3 without changing the overall behaviour of the equation too strongly. As a difference to ordinary polymerisation processes, evaporation may change the total weight fraction of monomer and

Table 4.2: Kinetic data.

	$A_i \,[\mathrm{m^3/(mol\,s)}]$	$E_{a,i}\,[\mathrm{J/mol}]$	$\Delta h_R\,[\mathrm{J/mol}]$
k_d	$\frac{\ln 1/2}{t_{1/2}} e^{\frac{E_{a,I}}{\Re T_{1/2}}}$	108×10^3	
	$t_{1/2} = 10\,\mathrm{h}, \quad T_{1/2} = 40\,^\circ\mathrm{C}$		
k_p	7.5×10^3	13×10^3	-77.5×10^3
k_t	47.5×10^6	15.464×10^3	
	$f_{td} = 0.05$		

polymer. Thus, the "initial" monomer concentration generally is not constant, but varies throughout the process. It needs to be evaluated in each time step depending on w_S, w_M and w_P. The modifications to k_p^0 and k_t^0 values obtained by Arrhenius equations are as follows:

$$k_p/k, p^0 = 1 + 14.8e^{-17w_M^0} \tag{4.1}$$

$$k_t/k_t^0 = e^{-X_P 360 w_M^{0,*3.7}} \tag{4.2}$$

$$w_M^{0,*} = \begin{cases} w_M^0 & 0 w_M^0 \leq 0.3 \\ 0.3 + 0.01\left(w_M^0 - 0.3\right) & 1 w_M^0 > 0.3 \end{cases}.$$

Basic 0D and 1D calculations have therefore been undertaken for both plain Arrhenius calculations and k_p and k_t modifications in order to examine the role of kinetics within the process. The investigation of process parameters by numerical design of experiments was conducted with modified k_p and k_t values.

Numerical experiments were carried out for an aqueous solution of acrylid acid as monomer. In order to accelerate radical formation and to limit the process time, VA-44 was chosen as initiator which has a very low 10 hour half-life decomposition temperature. The initial monomer content was 75 wt% and initiator was provided with a molar ratio to monomer of 2×10^{-4} if not stated otherwise. The initiator efficiency was set to 0.7, a reasonable value for free radical polymerisation. Under the assumption that no polymerisation reactions had taken place before and during atomisation, the initial polymer content was zero. The drying gas temperature was 95 °C at a relative humidity of 0 %.

The initial drop temperature was set to 20 °C. Heat and mass transfer were specified for constant drying gas properties at Nusselt and Sherwood numbers of 3. Partial pressures of surficial mixture components were calculated using UNIFAC thermodynamics. The initial droplet radius R_0 was set to 50 µm. Questions of initiator solubility after solvent evaporation were not regarded further in this study. The droplet's energy balance equation accounts for heat conduction, heat transport to the surrounding gas, the heat of evaporation and the heat of propagation reactions.

Distributed simulations have been carried out for constant and variable diffusion coefficients. Correlations for solvent-polymer diffusion coefficients are often based on the free volume theory (Duda et al. 1982; Vrentas, Duda, and Ling 1985; Vrentas, Duda, Ling, and Hou 1985) or experimental correlations.

In order to keep the diffusion approach simple, flexible and easily adjustable, a formulation based on the following rules was constructed:

- The diffusion coefficient decreases for higher polymer weight fractions w_P with a third power law: $D \sim (1 - w_P^*)^3$.

- The critical weight fraction, at which this power law approaches zero, is set as w_P^{crit}: $w_P^* = w_P/w_P^{crit}$.

- The decrease of the diffusion coefficient is limited to a certain order of magnitude $D_{log_{10}}^{lim}$.

The final modification of the diffusion coefficient is

$$\frac{D}{D^0} = \max\left(\left(1 - \frac{w_P}{w_P^{crit}}\right)^3, 0\right) + 10^{-D_{log_{10}}^{lim}}\left(1 - e^{-5w_P/w_P^{crit}}\right), \qquad (4.3)$$

in which D^0 is the initial diffusion coefficient in infinite dilution of polymer. The max function prevents the power law from becoming negative. Addition of both terms provides a comparably smooth transition between power law calculation and the constant limit diffusion coefficient. The factor $1 - e^{-5w_P/w_P^{crit}}$ prevents the sum of both terms from becoming larger than one at small w_P. Exemplary diffusion data for various parameter settings is plotted in Figure 4.1.

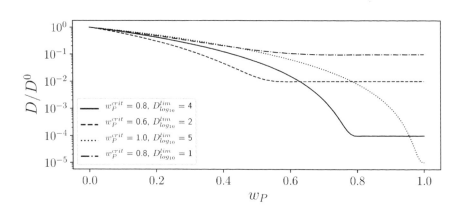

Figure 4.1: Dependence of the diffusion coefficient on the polymer weight fraction for different parameter settings.

Table 4.3: Antoine parameters
$(T \; [°C], p \; [mmHg])$

	water	AA
A	8.07131	8.68508
B	1730.63	2409.29
C	233.462	274.87

Acrylic acid has a boiling temperature of 141 °C at atmospheric pressure and is volatile under spray polymerisation process conditions. In order to examine the principles of droplet polymerisation and different effects independently from material properties, calculations were performed without and with monomer evaporation. Antoine parameters for water and acrylic acid are provided in Table 4.3.

4.2 Lumped Simulation of Droplet Polymerisation

4.2.1 Principle Course of the Process - Plain Kinetics, no Monomer Evaporation

As long as diffusion plays a minor role, a 0D approach describes spray polymerisation thoroughly and efficiently. Figure 4.2 shows the course of the process over 300 s. Plain kinetics without additional dependencies on initial monomer content or conversion have been applied and monomer evaporation was not considered. Immediately after drop formation, drying of the solvent takes place followed by a long phase of polymerisation. Hence, droplet shrinkage is very rapid during drying and thereafter more gradual due to the density increasing reaction. At the beginning of polymerisation, the temperature exhibits a small overshoot due to high conversion rates, but throughout the rest of the reaction time it remains virtually identical to the gas temperature of 95 °C.

A deeper look into the interplay of drying and polymerisation is provided in Figure 4.3, which contains the first two seconds of the process. The solvent evaporates within 0.5 seconds. Due to the heat of evaporation, the droplet is effectively cooled and remains at a temperature between 35 and 40 °C within this initial period. With decreasing solvent content, the concentrations of other components increase. After complete evaporation, the droplet is rapidly heated up and chemical reactions set in. The heat of reaction is mostly dissipated into the surrounding gas so that the droplet temperature rises only slightly above the gas temperature. Polymer is only produced after solvent evaporation which leads to two insights. The assumed concurrence of chemical reactions and evaporation (Biedasek 2009, p. 15) does not exist. In fact, solvent vaporisation and polymeri-

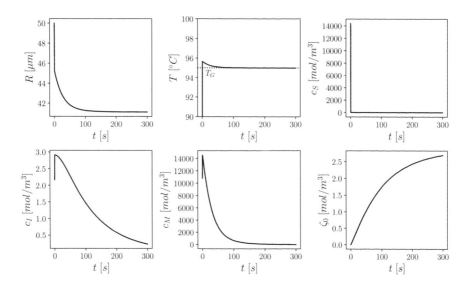

Figure 4.2: 0D simulation of polymerisation in a droplet over 300 s

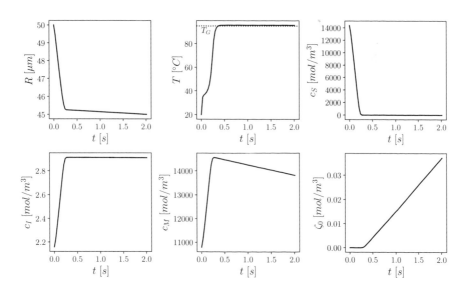

Figure 4.3: 0D simulation of droplet polymerisation within the first two seconds.

sation are two consecutive processes and the second one starts not before the first one is finished. Secondly, polymerisation takes place as bulk/mass polymerisation but not within a solution. Both findings depend on the assumption that the 0D approach sufficiently describes spray polymerisation, i.e. solvent evaporation is not hindered by diffusion. This will be further discussed by means of the 1D model in section 4.3. Subsequently (Figure 4.2), polymer of a higher density than the monomer is formed so that the droplet shrinks further. The heat of reaction is efficiently removed from the droplet due to its beneficial surface area to volume ratio. Whereas nearly full conversion is achieved at about 200 s, the zeroth moment still rises significantly and has not approached its final value after 300 s. Continuous initiator decomposition at a minimal rest of monomer leads to ongoing formation of short polymer chains.

4.2.2 Effects of Kinetics on the Process

The simple introductory example exhibited an increase of polymer concentration even at practically full conversion and a rather long process time compared with the residence time in a spray dryer.

Initiator Efficiency

The first phenomenon is due to the fact that chain propagation and initiation evolve differently over time. The ratio of initiator to monomer rises and leads to formation of short chains when only a small monomer fraction is left in the droplet. This becomes clearer by the moments of dead chains in Figure 4.4 (solid lines). Whereas the zeroth moment increases even at very low monomer contents, the first moment (= the total amount of monomer units in polymer chains) changes only slightly after 100 s and remains virtually unaltered after 200 s. The same holds for higher moments. Thus, the number average of the chain length distribution P_n decreases with the additional build-up of small chains while the weight average P_w approaches its limit value earlier. As a result, the dispersity $Ð = P_w/P_n$ rises strongly and differs significantly from the typical value for radical polymerisation of 2. Chain initiation - build-up of iniator radicals and their reaction with monomer to chain radicals - is implemented as one single reaction with initiator decomposition being the rate determining step. This is valid under the assumption of monomer abundance and allows for not considering short-living initiator radicals as an additional component. The initiator efficiency specifies how many chain radicals evolve on average by decomposition of one initiator molecule. At low monomer contents it becomes more unlikely that an initiator radical approaches a monomer molecule instead of terminating itself by other reactions before. This was additionally modelled by a simple approach for a diminished initiator efficiency at small monomer concentrations

$$f_d = f_d^0 \left(1 - e^{c_M/c_M^{I63}}\right). \tag{4.4}$$

c_M^{I63} is the monomer concentration of 63 % of initial initiator efficiency f_d^0. If this value is higher, initiator efficiency will go down earlier. For the dashed

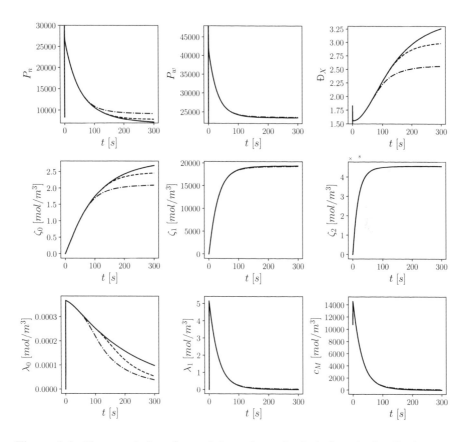

Figure 4.4: Characteristic values of the polymer's chain length distribution and statistical moments, 63 % initial initiator efficiency at 1 (solid line), 100 (dashed) and 500 mol/m³ (dash-dotted) monomer concentration.

and the dash-dotted lines in Figure 4.4, c_M^{I63} was set to 100 and 500 mol/m^3, respectively. As a result, the concentration of living chains λ_0 decreases earlier and the total number of polymer chains ζ_0 approaches its plateau sooner and at lower final values. The higher moments of living and dead chains remain unaltered. Hence, the number average remains at higher values and the dispersity does not increase that strongly. As maximum monomer concentrations in the bulk are about 14 500 mol/m^3, around one percent of monomer is affected by $c_M^{I63} = 100$ mol/m^3 with the strongest effect on a few permille.

Trommsdorff-Norrish Effect

With a larger amount of polymer being created, the viscosity of the solution increases and the mobility of chains is lowered. The probability of two living chains encountering each other becomes lower and termination more unlikely. This typical Trommsdorff-Norrish or gel effect was modelled in Figure 4.5 by the conversion dependency of equation 4.2. The solid line shows the process without, the dashed line with the gel effect ($c_M^{I63} = 100$ mol/m^3 in both cases). The onset of the gel effect takes place after about 5 s. Due to the strongly increased conversion rate, the excess temperature is slightly higher. Still, the con-

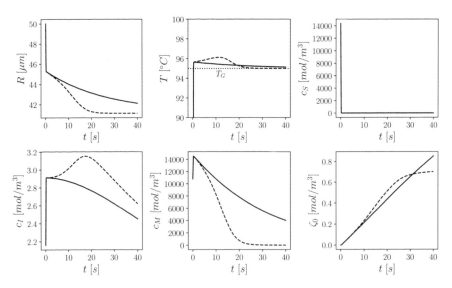

Figure 4.5: 0D simulation of polymerisation without (solid) and with the Trommsdorf-Norrish effect included (dashed).

trol of the reaction temperature is very good due to the droplet's large ratio of surface area to volume. As the drop shrinkage by cause of density changing reactions is accelerated, the inititaor concentration further rises after the evaporation period until monomer conversion becomes less intense. Moreover, the increase of polymer concentration at very low monomer contents is much less pronounced because termination of these small chains becomes unlikely and monomer is rather consumed by propagation. The overall process time is nearly an order of magnitude smaller than without taking the gel effect into account.

Figure 4.6 shows the corresponding course of conversion, zeroth moments of living and dead chains and characteristic values of the chain length distribution. Additionally, data for the gel effect with practically constant initiator efficiency ($c_M^{I63} = 1\,\text{mol/m}^3$, dash-dotted line) is included. Inclusion of the gel effect induces full conversion after 25 s, independently from the initiator efficiency at low monomer contents which only affects the last remaing monomer molecules. With termination becoming improbable due to the gel effect, the living chains' concentration λ_0 rises rapidly until the time of (nearly) full conversion. At constant initiator efficiency, λ_0 remains on this high level until virtually the last

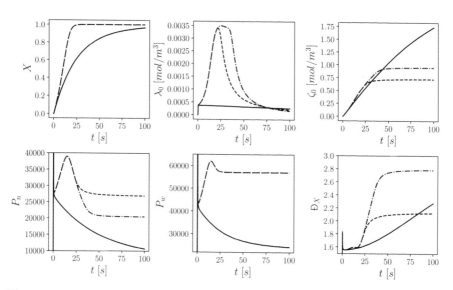

Figure 4.6: Conversion, concentrations of polymer chains and characteristic values of the CLD calculated without (solid line) and with the Trommsdorff-Norrish effect for $c_M^{I63} = 100\,\text{mol/m}^3$ (dashed) and $1\,\text{mol/m}^3$ (dash-dotted).

monomer molecules are consumed. Otherwise it decreases strongly afterwards, as less chains are initiated ($c_M^{I63} = 100\,\mathrm{mol/m^3}$). Consequently, the dead chains' concentration ζ_0 approaches its plateau earlier and at a smaller value in the latter case. The gel effect involves a higher number average of the chain length distribution as the ratio of propagation to termination is much higher. Again, an earlier decrease of initiator efficiency leads to significantly higher P_n values, whereas higher moments and the weight average are independent of the initiator model. With the Trommsdorf-Norrish effect included and $c_M^{I63} = 100\,\mathrm{mol/m^3}$, the dispersity changes only to a minor degree and finally remains slightly above the typical value for free radical polymerisation of 2.

The implementations of initiator efficiency, the Trommsdorf-Norrish effect and propagation dependency on monomer content are simple, but show the impact of kinetic effects on polymerisation within a drop and on the overall process time. Subsequent calculations include the gel effect and c_M^{I63} is set to $100\,\mathrm{mol/m^3}$. In doing so, studies can be evaluated without the need of addressing side effects which alter results at very low concentrations. Whilst these values do not appear completely unrealistic, it has to be born in mind that the kinetic model is simple and only partly based on measurements. Under the necessity of fast conversion in spray polymerisation, kinetic parameters are needed for process conditions that are not commonly covered by experiments. Hence, the following studies will concentrate on qualitative features of the process.

4.3 Spatial Effects in Droplet Polymerisation

Section 4.2 showed that kinetic parameters strongly affect a spray polymerisation process. Due to drying, inhomogeneities may occur within the process when the surficial solvent concentration is low and still near its feed value within the droplet's core. This also affects the monomer and initiator concentrations and can therefore lead to an inhomogeneous product. In the following, the results of a 1D model as proposed in section 3.3.4 will be compared with the corresponding lumped simulations for various cases. Diffusion was modelled by the pseudo-binary Fickian approach provided at the end of section 3.1.5. If not stated differently, simulations have been carried out with the Trommsdorf-Norrish effect and k_p dependency on w_M^0 being active. Feed monomer content was set to 75 wt%. Monomer evaporation was switched off in the simulations considering effects of diffusion.

4.3.1 Effect of the Diffusion Coefficient on Concentration Gradients

A typical order of magnitude of diffusion coefficients in an aqueous solution is $D = 1 \times 10^{-9}\,\mathrm{m^2/s}$. This value was used for the 1D simulation in Figure 4.7. Concentrations are provided as radial profiles at intervals of $4\,\mathrm{s}$. The related 0D results are depicted as circles for the same instants of time and in between as dashed lines. The instants of the radial profiles are indicated in the time-dependent R and T graphs additionally by circles and dashes for the lumped and distributed simulation, respectively. Initial profiles are drawn in bold.

Both calculations are practically identical. The profiles are flat and 0D results are matched over the whole droplet. Only minor differences occur within the first two seconds of the process as is depicted in Figure 4.8 in which averaged 1D values are compared with their 0D counterparts. At low solid contents the drying rate is overestimated in the 0D model, in which, due to the abscence of radial gradients, the surficial solvent content and partial pressure remain at a

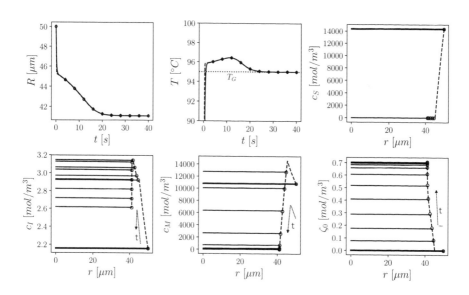

Figure 4.7: 1D simulation of polymerisation in a droplet with a diffusion coefficient of $D = 1 \times 10^{-9}\,\mathrm{m^2/s}$, radial profiles at $0, 4, 8, ..., 40\,\mathrm{s}$, corresponding 0D results as circles and dashed lines.

higher level until solvent evaporation is finished. Its behaviour is therefore more stepwise, whereas the transition between drying at small droplet temperatures with high shrinking rates to chemical reactions at high temperatures is more gradual in the distributed simulation. As full solvent evaporation takes about 0.7 s longer in the 1D model, polymerisation is slightly retarded as well. Hence, polymer concentrations are marginally smaller and the droplet radii a bit higher compared to the 0D simulation at same instants of time. Still, the differences are so small that after two seconds both results can be regarded as equivalent.

This result may be surprising at first glance as the mass transfer Biot number is one order of magnitude higher than its limit value which allows for lumped modelling (section 2.5.4). Yet, these results are not to be mistaken. The mass transfer Biot number refers to solvent diffusion in the droplet. In fact, the course of solvent concentration over time is significantly different in lumped and distributed modelling. The time until full evaporation differs by a factor of about three. If solely solvent vaporisation was of interest, the error introduced by lumped modelling would indeed be very high. Yet, considering that polymerisation sets in just after evaporation and that the reaction product is of main interest, it is irrelevant whether full solvent evaporation was achieved after less than 0.4

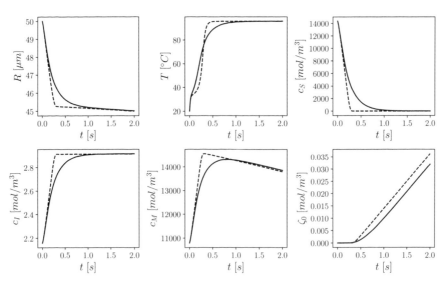

Figure 4.8: Comparison of 0D (dashed lines) and averaged 1D results (solid, $D = 1 \times 10^{-9}\,\mathrm{m^2/s}$) within the first two seconds of spray polymerisation.

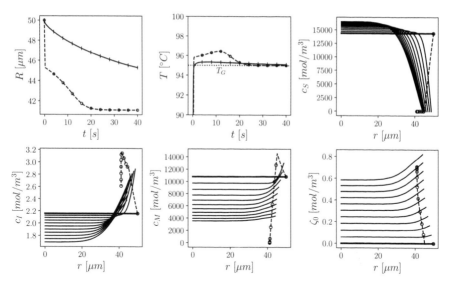

Figure 4.9: Comparison of 0D (dashed lines, circles) and 1D results (solid) for a small diffusion coefficient of $D = 1 \times 10^{-12}\,\mathrm{m^2/s}$ at $0, 4, 8, \dots, 40\,\mathrm{s}$

or slightly above 1 s. Hence, lumped and distributed modelling provide virtually equal results of the reactive drying process despite the strong relative deviation in solvent evaporation at the process' beginning.

Figure 4.9 provides a 1D simulation for $D = 1 \times 10^{-12}\,\mathrm{m^2/s}$, a diffusion coefficient which can be considered as reasonable for a polymer rich solution. In this case, strong radial concentration gradients occur. Directly after the droplet is exposed to the drying gas, the solvent concentration at the droplet's surface decreases within fractions of a second. Afterwards, evaporation is limited by transport from the droplet's core to its surface. The quasi-stationary equilibrium of evaporation and diffusive solvent transport from the core involves a small solvent content at the surface and low drying rates. Hence, the drop is heated up almost immediately. Chemical reactions set in, but - as initiator concentrations are lower and therefore less chain radicals are induced - at lower rates compared to the case with no diffusion limitation. Due to this and additional slight cooling by solvent evaporation, the temperature overshoot is smaller compared to the 0D simulation. Whereas the solvent concentration exhibits strong gradients over the drop's radius, this is less pronounced for the initiator. The monomer concentration nearly even flattens out within the simulated 40 s. This is not only caused by diffusion of monomer and initiator to the droplet's core, but also by chemical reactions which are faster at the surface due to the high educt concentration. Initiator decomposition is accelerated and a higher number of polymer radicals created. Monomer conversion is hence accelerated by a higher educt concentration of both living chains and monomer itself. Therefore, the surficial monomer content decreases even faster. The higher educt concentrations near the droplet's surface result in a larger concentration of polymer.

4.3.2 Inhomogeneities of the Product at Small Diffusion Coefficients, Effect of Moments' Diffusion

Polymer properties of the previous simulation with $D = 1 \times 10^{-12}\,\mathrm{m^2/s}$ are shown in Figure 4.10. The upper row contains number and weight average and dispersity profiles, when just Arrhenius equations were used for calculation of reaction parameters. Despite the strong gradients of reactants at the beginning of the simulation, the number and weight average of the chain length distribution vary only slightly over the droplet radius. Spatial gradients are much smaller than changes over time. The resulting product can still be considered as fairly homogeneous. In the middle row, additional conversion and w_M^0 dependency of termination and propagation according to equations 4.1 and 4.2 had been switched on. For one thing, the changes over time are stronger, with P_n and P_w running through a maximum near 16 s. This is in accordance with 0D calculations including the Trommsdorf-Norrish effect (compare Figure 4.6). For another thing, spatial variations are somewhat higher as well, but still distinctly lower than differences throughout the process time. For reasons of clarity the final P_n and P_w profiles at 40 s are plotted in subgraphs. Values at the outer shell are about 10 % higher compared to the inner core. The picture changes, when the feed monomer content is reduced from 75 to 25 wt%. In this case, spatial differences of the monomer concentration between droplet core and outer shell are much higher and result in large inhomogeneities of number and weight average of the chain length distribution. It has to be kept in mind that a sphere's volume grows with the third power of the radius. When deviations, as in the depicted case, only affect the outer 20 % of P_n and P_w profiles, this outer shell contains about half of the droplet volume though.

The same calculation as in the lower row, but with simplified Fickian diffusion acting independently on all moments (cmp. equation 3.36), is shown in Figure 4.11. As before, including k_p dependency on w_M^0 involves shorter chains at the droplet's outer rim. Yet, profiles are smeared out so that these inhomogeneities are not as distinct as when modelled correctly. Moreover, due to unrealistic polymer transport, higher chain lengths are predicted even at intermediate positions in the droplet (e.g. 20 µm), at which solution concentrations are mostly unaffected by drying and similar to those at the drop's centre. As proposed theoretically before, a simple model of Fickian moment diffusion will, given enough time for equilibration, always end up in homogeneous product features. A proper polymer diffusion model is therefore necessary, when diffusive

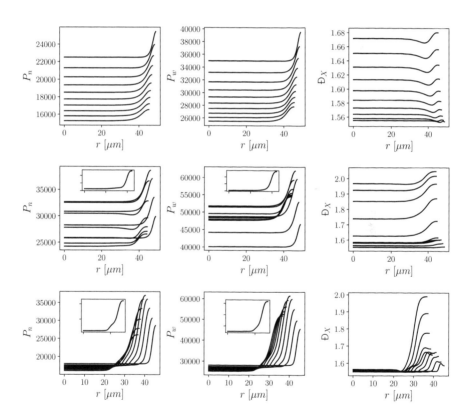

Figure 4.10: Characteristic values of the polymer for $D = 1 \times 10^{-12}\,\mathrm{m^2/s}$, upper row: simple second order kinetics, middle row: dependency on w_M^0 and X_P taken into account, lower row: as middle row, but with $w_M^0 25\,\mathrm{wt\%}$; profiles at $4, 8, \ldots, 40\,\mathrm{s}$; small subplots show final profiles at $40\,\mathrm{s}$.

transport of moments is to be considered.

The assumption of a very low diffusion coefficient is valid when a considerable amount of polymer is present. Still, the polymer content is virtually zero at the beginning of the spray polymerisation process so that the diffusivity is similar to the one of an ordinary solution. With further progress of polymer formation, diffusion will be hindered increasingly. In Figure 4.12, the diffusion coefficient was varied according to equation 4.3 with $w_P^{crit} = 0.8$ and $D_{log_{10}}^{lim} = 4$. In doing so, the diffusion coefficient will be lowered by four orders of magnitude for a polymer weight fraction beyond 0.8, whereas it remains unaltered in abscence of polymer. The initial diffusion coefficient D^0 was set to $1 \times 10^{-9}\,m^2/s$. Flat profiles are obtained similar to the results of an unmodified, constant diffusion coefficient D^0. Considering again the course of process within the initial seconds in Figure 4.8, it becomes clear that as long as solvent evaporates, the droplet is strongly cooled and polymer formation is effectively prevented. As long as there is no polymer built up, the diffusivity stays at a high level so that drying is not limited by transport within the droplet. Hence, the solvent completely evaporates before a considerable amount of polymer is formed and diffusion can be hindered in fact.

Interim Conclusion

Considering the test case with solely solvent evaporation and low humidity in the drying gas, it is clear that spatial gradients will only occur when polymer is already present within the droplet at the beginning of the process. The the-

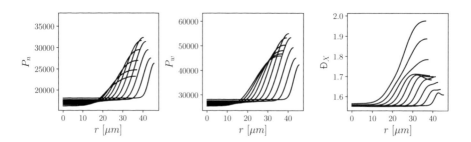

Figure 4.11: Number and weight average and dispersity of the chain length distribution at $4, 8, ..., 40\,s$ for $D = 1 \times 10^{-12}\,m^2/s, w_M^0 = 25\,\%$, simple Fickian diffusion acting independently on all moments.

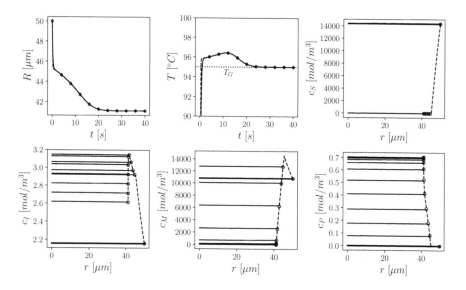

Figure 4.12: 0D (dashed lines, circles) and 1D results (solid) for a diffusion coefficient varying between 1×10^{-9} and $1 \times 10^{-13}\,\mathrm{m^2/s}$, depending on the polymer content, profiles at $0, 4, 8, ..., 40\,\mathrm{s}$.

oretical case of a very large droplet, in which diffusion would be too slow for equaling out gradients even at $D = 1 \times 10^{-9}\,\mathrm{m^2/s}$, is not applicable to spray processes and only of academic nature.

As chemical reactions are efficiently prevented in presence of solvent due to the heat of evaporation, the process can be divided into two different phases, a period of sole evaporation and the stage of synthesis, which is then carried out as bulk polymerisation.

As long as only solvent evaporation is occuring and no other effects evoke a strong decrease of the diffusion coefficient, a lumped model represents the droplet polymerisation process as good as a distributed approach. In terms of reaction modelling this allows for the implementation of more complicated reaction mechanisms for which implementations in a 1D model become costly. Such schemes could involve moments of tertiary radicals, copolymers or the like or even a detailed simulation of the molar weight distribution with algorithms such as adaptive h-p Galerkin methods (Wulkow 2008).

4.3.3   Effect of Monomer Evaporation

So far monomer vaporisation has been neglected in order to study the model behaviour in case of hardly evaporating monomers. Yet, acrylic acid is a volatile
component and the assumption of sole solvent evaporation is an unsuitable simplification. As Figure 4.13 shows, the process changes drastically, when the
monomer is also affected by drying. Same as for the solvent, the relative saturation of the drying gas with respect to monomer was set to 0% and a variable
diffusion coefficient starting at $1 \times 10^{-9}\,\mathrm{m^2/s}$ was assumed as before. Within
the first $0.5\,\mathrm{s}$ the process is mostly unchanged to the previous example. The
solvent concentration at the droplet's surface decreases strongly due to evaporation. Under abscence of polymer, the diffusion coefficient remains at its original
value so that diffusive transport from the drop's bulk to its surface is scarcely
hindered. The monomer concentration, like the initiator's one, rises at the droplet's surface at first. After most of the solvent was vaporised and its surface
concentration gets low, the activity of monomer at the interface to the drying
gas is increased due to its higher concentration and monomer evaporation starts.
At the same time, solvent evaporation and as a consequence thereof cooling of
the droplet are weakened. The drop's temperature rises to a new quasi-steady-
state, at which heat transfer from the drying gas and cooling due to monomer
evaporation are in equilibrium. This is similar to solvent vaporisation, but at an
elevated temperature so that polymerisation reactions partly take place. Finally,
a small sphere remains consisting of polymer and a considerably high fraction
of initiator.

Comparing 0D and 1D results, the lumped model exhibits a more stepwise
behaviour again. Solvent is evaporated nearly completely before monomer evaporation sets in. In the 1D case the solvent concentration in the droplet's core
remains at a higher value when monomer vaporisation already takes place. The
temperature curve of the 0D model shows a clear distinction between evaporation of solvent and monomer, whereas the transition between these two phases is
gradual in the distributed case. Nevertheless, the outcome of both models is virtually the same until polymerisation begins to play a role, especially comparing
surface values of the 1D calculation with 0D results. However, polymerisation
reactions lead to a different result in the distributed simulation. The final radius
is a bit higher than in the 0D case. At some point, evaporation of monomer is
hindered and a slightly higher number of monomer molecules is converted into

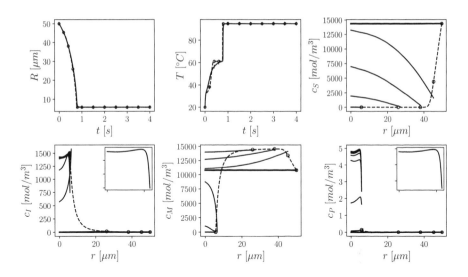

Figure 4.13: 1D and 0D simulation of AA spray polymerisation with evaporation of the monomer at $0, 0.4, 0.8, ..., 4\,\mathrm{s}$; subgraphs show final profiles of initiator and polymer concentration.

polymer. Despite the difference being small, it is large enough that the initiator concentration approaches a lower value. Moreover, the polymer concentration is drastically higher. Final profiles of these two components are depicted in the subgraphs. Interestingly, the concentrations of both remaining species exhibit a distinct minimum at the droplet surface. This is not a violation of conservation laws, as it might look at first sight, but an effect of the polymer component with locally varying molecular weight and a maximal mole specific volume at the droplet surface. Characteristic values of the chain length distribution are depicted in Figure 4.14. Final profiles at $4\,\mathrm{s}$ are again drawn in subplots. At the end of the calculation, P_n is about a factor of five higher at the surface than in the droplet's core. Despite the low concentration of polymer molecules, the amount of monomer units in chains and the volume occupied by polymer are highest at the droplet surface.

The difference in the calculated number average between 1D and 0D calculations is striking. During the phase of mainly monomer evaporation both models calculate similar P_n values. Whereas the 0D model remains at this high level, the number average decreases rapidly in the distributed case with final values being

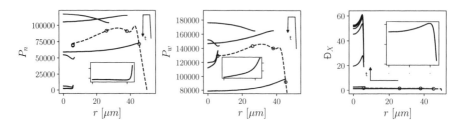

Figure 4.14: Characteristic values of the chain length distribution for polymerisation with evaporation of the monomer at $0.4, 0.8, ..., 4\,\mathrm{s}$; subgraphs show final profiles

more than one order of magnitude lower. A similar behaviour can be observed for the weight average, yet to a much smaller degree and with its surficial value remaining at the same level as the corresponding 0D value. With both values evolving differently, the dispersity of the product grows strongly to an average value above 50. Figure 4.15 provides more insight into this surprising outcome of the 1D simulation. Concentration, moments and characteristic values of the chain length distribution are plotted over the droplet temperature. 1D results (solid lines) have been condensed into averaged values.

As discussed before, the 0D model exhibits a stepwise behaviour. In the very first phase the droplet is heated up until cooling by solvent vaporisation balances heat transfer from the surrounding gas at about 35 °C. Due to solvent evaporisation, the monomer concentration rises. It comes to its maximum, when virtually no solvent is left and monomer evaporation becomes dominant. A second equilibrium of cooling due to the heat of monomer vaporisation and heat transfer from the surrounding occurs slightly above 60 °C. This temperature level is larger than in the preceding phase, because the vapour pressure and the heat of evaporation of acrylic acid are smaller compared to water. Now, monomer evaporates and the concentration of the remaining initiator rises rapidly. At the same time, polymerisation reactions set in so that a small portion of monomer is converted into polymer. This period of concurrent chemical reactions and to a much larger degree evaporation takes place in a quasi-steady-state at a mostly constant temperature. As can be observed from the moments ζ_0 and ζ_1, virtually all polymer molecules are build up in this phase. P_n and P_w values and hence the dispersity remain at the same level, when evaporation and reactions break down

108

and the droplet is finally heated to the surrounding gas temperature.

In the 1D model, by way of contrast, these processes are superimposed by diffusion effects. As was already discussed concerning Figure 4.13, the concentration profiles exhibit strong gradients. The surficial solvent fraction decreases earlier so that the monomer content is elevated there. Hence, the temperature evolution during this quasi-steady-state period of solvent drying is more smeared out and does not remain near a certain level as in the 0D case. The two phases of solvent and monomer evaporation are not separated that distinctly as in the lumped model. Monomer is already vaporised before the solvent is completely consumed. Same as in the 0D calculation, polymerisation reactions

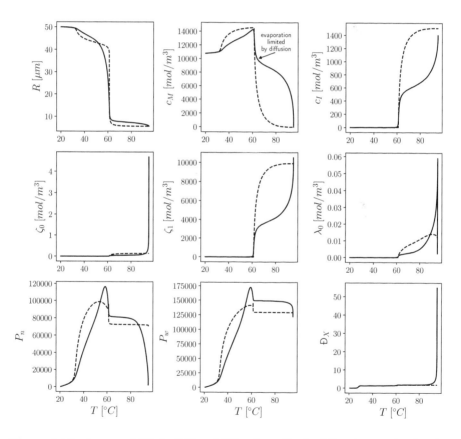

Figure 4.15: Averaged 1D (solid) and 0D values (dashed lines) over the droplet temperature for the case of droplet polymerisation with monomer evaporation.

set in during this phase and the course of the first moment over the temperature is identical until it exceeds a concentration of $2000\,\text{mol}/\text{m}^3$. At this point, when the droplet radius is already smaller than $10\,\mu\text{m}$, the diffusion coefficient at the droplet's surface has decreases due to the growing polymer fraction that much that transport of monomer from the drop's core is slowed down. The surficial monomer concentration and its evaporation rate are lowered further. Within a short transition time, the droplet heats up and chemical reactions are accelerated while evaporation still is partly taking place. When diffusion is hindered that much that drying is practically stopped, the drop temperature approaches the level of the drying gas temperature and the rest of monomer is converted into polymer. Initiator decomposition is accelerated strongly at this higher temperature. Hence, a large amount of chain radicals is created (see the zeroth moment of the living chains' distribution λ_0). Comparing the values at which most of the polymer is created - $> 90\,°\text{C}$ in the 1D and $\sim 60\,°\text{C}$ in the 0D case - the living chains' concentration is about one order of magnitude higher in the distributed simulation. The initiator to monomer ratio is exceedingly high with values above $1:10$. Therefore, a large number of very short chains is added in this final reaction period resulting in a tremendous drop of the number average P_n. Higher moments are not affected so strongly, as for instance the total amount of monomer units - the first moment - remains in the same order of magnitude. The effect on the weight average is therefore much smaller. As a result, the dispersity grows to unusually high values.

Comparing the final droplet radii without and with monomer evaporation of about 45 and below $10\,\mu\text{m}$, less than one hundredth of the initial monomer content remains in the drop. In reality, the effects may be less or even more pronounced than being calculated in the 1D model, depending on the relationship of diffusivity to the solution's composition. Provided that monomer evaporation is scarcely hindered by polymer formation, the result will be as predicted in the 0D calculation. If diffusion is hindered earlier than presupposed here, more monomer will remain within the droplet. The initiator to monomer ratio will still approach unuasally high levels, yet smaller than in the present calculation.

Interim Conclusion

Clearly, only a distributed model covers all relevant processes when educt concentrations within the droplet are affected by diffusion. Monomer evaporation plays a major role concerning the yield of droplet polymerisation. The process is unattractive in such a regime, even, if a very efficient recovery of monomer from the drying gas could be applied. Moreover, polymer properties are affected due to the large change in reaction conditions.

As a remedy, transport of monomer to the drying gas has to be limited. This can either be achieved on the droplet side by preventing transport to the gas interface or on the gas side by reduction of the driving force for evaporation. The first approach can be realised by providing polymer already at atomisation of the drop so that the diffusivity drops to low levels at an earlier instant of time. The second solution may be achieved by milder drying conditions with respect to the monomer, for example by a higher monomer saturation in the drying gas. Both approaches will be discussed in the following two examples and further investigated by means of numerical DoEs in section 4.5. A third way could involve additional additives, which lower the activity of the monomer to such a degree that evaporation becomes subordinate.

4.3.4 Pre-polymerisation Before Atomisation

The previous example showed that only a small fraction of the initial monomer content may be converted into polymer if it is a volatile species. One solution is to provide polymer already within the feed material, which will build a skin of low diffusivity at the droplet surface under drying. A pre-polymerisation period was implemented into the model by switching off mass transfer to the surrounding and clamping the temperature to a level $T_{prepoly}$ within an initial process phase. In order to provide smooth transitions to the numerical solver, this was realised via a prefactor, gradually switching heat and mass transfer to the surrounding from zero to its full value(see also Figure 4.16, left plot):

$$f_{prepoly} = 0.5 + 0.5 \cdot \tanh(200t). \tag{4.5}$$

This factor was applied to mass fluxes in the boundary conditions at $r = R$ and also to the energy balance in order to keep the temperature at the same, controlled level. The real equivalent to this model setup could be feed material running through small tubes, in which the temperature control is good enough to prevent a runaway of reactions, and being atomised thereafter. Figure 4.17 shows a calculation including 5 s pre-polymerisation at the same temperature as in the drying gas. Concentrations before atomisation are drawn as flat profiles over the droplet radius in the 1D case. Despite no droplet was created at these instants of time, this representation makes it easier to evaluate the evolution throughout the process and to distinguish 1D and 0D results.

Due to the smooth numerical switch, heat and mass transfer to the surrounding are not instantly switched on at 0 s, but drying has already partly started before and just reached 50 % of its full effect by this point. Hence, the third radial profile in the distributed case as well as the respective circle from the 0D calculation already show the effect of mild drying. The lumped simulation is easily explained. Before atomisation, the process is just executed like in a batch or in a plug flow within a tube reactor. After exposition to the drying gas, the solvent evaporates first, just as in the previous example. After (nearly) complete solvent vaporisation, the monomer species is exposed to drying, which happens rapidly and consumes most of the remaining monomer. Concurrently, chemical reactions take place and small fractions of additional polymer are built. One difference to the previous case is that, starting at a the drying gas temperature, water evaporation takes place at higher temperature so that chemical reactions are slightly more in favour. Secondly, with polymer being present, the monomer fraction is smaller so that the drying period is reduced. The polymer created in the pre-atomisation phase is by far the main constituent of the final product in the 0D case. As the right graph in Figure 4.16 shows, the lumped model virtually simulates a polymerisation process in a tube reactor followed by spray drying.

Drying and chemical reactions are again superimposed by diffusion effects in the 1D simulation. The presence of polymer at the droplet's surface decreases the diffusivity so that solvent evaporation practically stops within a second. The

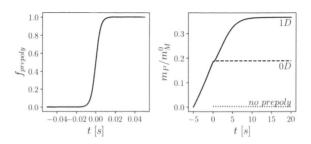

Figure 4.16: Left: numerical switch for transition between pre-polymerisation (0) and normal droplet polymerisation (1), right: polymer mass over time.

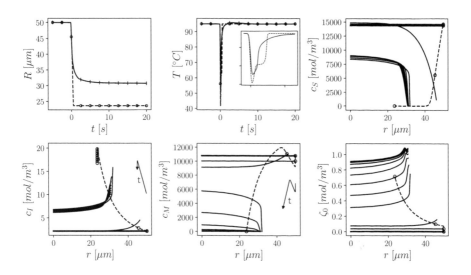

Figure 4.17: 1D and 0D results of droplet polymerisation with monomer evaporation and a 5 s period of pre-polymerisaiton before atomisation; profiles at $-5, -2.5, 0, 2.5, ..., 20$ s.

temperature drop due to evaporative cooling is less pronounced than in the 0D case. Likewise, subsequent and partly parallel monomer evaporation stops soon after. Under absence of solvent and monomer at the droplet's surface, the diffusion coefficient locally becomes very low. Hence, all species exhibit steep concentration gradients near the interface to the drying gas. Additional polymer production lowers the diffusion coefficient further within the droplet. Final concentration gradients are therefore a bit smoother, but still very pronounced as transport only scarcely takes place.

The final product exhibits a substantially larger radius than in the 0D calculation, not only as more monomer is converted into polymer, but also as still a large amount of solvent remains when transport to the drying gas is practically prohibited. The solvent concentration in the droplet's core even rises after evaporation has stopped, as the drop shrinks by cause of density changing reactions and the remaining solvent is concentrated within a smaller volume. Figure 4.16, right frame, shows that in the 1D case about twice as much monomer is converted into polymer as in the lumped simulation. Yet, more than half of the initial monomer still evaporates into the drying gas.

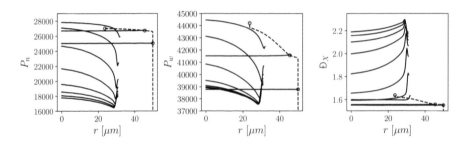

Figure 4.18: Characteristic values of the chain length distribution for the pre-polymerisation case at $-2.5, 0, 2.5, ..., 20\,$s.

The characteristic values of the chain length distribution are plotted in Figure 4.18. Generally, the changes over time are much smaller compared to the previous example without pre-polymerisation (cmp. the y-axis scale). In the 0D calculation, the largest part of the final polymer is created before atomisation under nearly constant conditions so that chain length distribution properties change only little. As single profiles are hard to identify in Figures 4.17 and 4.18, profiles at 0.5 (dashed) and the final state of $20\,$s (solid line) are drawn in Figure 4.19. After $0.5\,$s the surficial polymer weight fraction is already nearly one. The diffusion coefficient thus exhibits a sharp drop near the droplet's surface. As only a small outer rim of the droplet is affected, diffusive transport can still take place under locally high mole fraction gradients. Yet, it is already strongly limited. With drying being very weak at this instant of time (compare Figure 4.17), the profile is not a snapshot within a very transient phase of strong changes but represents a quasi-steady-state despite its steep gradient at the surface and the edge in the profile. Both peculiarities just correspond with the strong local variations of the diffusion coefficient and provide a continuous course of the diffusive flux at this point. Throughout the process, mole fraction profiles are smoothed, but the steep gradients at the droplet's surface largely remain.

The final solvent content is partly reduced and its profile is smoothed. The product consists of more than $80\,$wt% polymer everywhere in the droplet. The diffusion coefficient therefore approached its lower limit value of $1 \times 10^{-13}\,\mathrm{m^2/s}$ throughout the whole product. Considering the time scale of a spray process the final solvent content is "frozen". The time span of full evaporation shifts from less than a second to more than an hour.

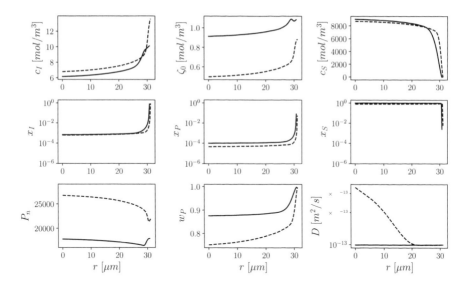

Figure 4.19: Radial profiles of the pre-polymerisation case at 0.5 (dashed) and 20 s (solid lines).

Interim Conclusion

Pre-polymerisation as a process variant allows for a reduced loss of volatile monomers to the drying gas. Moreover, the residence time for creation of the same amount of polymer per droplet can be reduced, as the process is partly shifted prior to atomisation. On the downside, removal of solvent is strongly exacerbated. Technically, this process variant involves additional issues such as heat control in the pre-polymerisation phase or atomisation of a polymer rich solution and clogging. Moreover, it not only prevents monomer evaporation, but also solvent vaporisation. The amount of generated polymer before atomisation therefore needs to be adjusted carefully.

4.3.5 Polymerisation at Elevated Monomer Content in the Drying Gas

A higher partial pressure of monomer in the drying gas will reduce the driving force for monomer evaporation. In the following example, monomer saturation was set to 90%. Other drying and feed conditions were kept at the same levels as in previous simulations. Figure 4.20 shows the evolution of droplet radius and temperature and solvent, initiator, monomer and polymer concentrations during the first 7.5 s. Comparison with Figure 4.12 shows that, same as before, solvent evaporation takes place almost instantaneously and the monomer concentration rises to a similar level. However, the increase in monomer content happens not only due to reduction of solvent but also by monomer uptake from the drying gas. The droplet radius rises despite solvent evaporation and, different to previous simulations, the initiator concentration is lowered. As long as the droplet temperature stays below the dew point of the monomer (92.4 °C under the present conditions), condensation of monomer from the drying gas at the "cold" droplet will take place. If the droplet consisted of pure monomer, uptake from

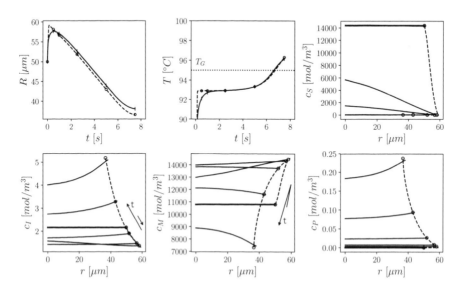

Figure 4.20: Droplet polymerisation with 90% monomer saturation in the drying gas at $0, 0.5, 1, 3, 5, 7.5$ s.

the drying gas would stop as soon as the dew point was surpassed. In a mixture, the monomer activity typically will be lowered. Hence, its partial pressure at the droplet surface may be considerably smaller than in the surrounding gas. As a result, the droplet serves as an absorber for monomer from the drying gas. Both effects lead to a monomer uptake within the first second of the process.

Thereafter, the temperature at the drop's surface is so high that monomer evaporation sets in. Other than before, the drying conditions are mild. Within the next seven seconds, the droplet shrinks to about 75 % of its initial radius, as evaporation is strongly limited and also buffered by the amount of monomer additionally taken up within the first second. In comparison, it takes less than half of a second under harsh conditions to vaporise most of the monomer (see Figure 4.13). After this period of time, the monomer content at the droplet's surface has become so low that evaporation stops. The course of the drop radius therefore approaches a flat tangent at the end of this phase.

The temperature level during this quasi-steady-state period is comparably high, starting slightly below the drying gas temperature and surpassing it when monomer evaporation stops. Therefore, chemical reactions take place throughout this phase and a considerable amount of polymer is created. This is also the reason for stopping monomer evaporation after about ten seconds, as the polymer content at the surface becomes so high that monomer activity is substantially lowered. Comparison of lumped and distributed results shows that surficial values of the 1D simulation are almost identical to corresponding 0D values. Inside the droplet, creation of polymer partly hinders diffusion so that even under mild drying conditions monomer transport to the outer rim is decreased. Hence, the monomer concentration is higher and the initiator concentration lower inside the droplet. For the same reason, the droplet radius is a little higher in the distributed case.

Figure 4.21 depicts the course of the process starting from 10 s (bold profiles) until 120 s. First, 0D simulation shall be discussed. After ten seconds of the process, the monomer content has decreased so much that its partial pressure at the drop's surface becomes lower than in the surrounding gas. The droplet begins to take up monomer from the drying gas. In the simple, lumped approximation, the process approaches a new quasi-steady-state of reactive absorption, in which the (fully mixed) monomer concentration stays at a nearly constant level and the amount of absorbed monomer is directly transformed into polymer. With decreasing initiator content - due to chemical reactions and dilution

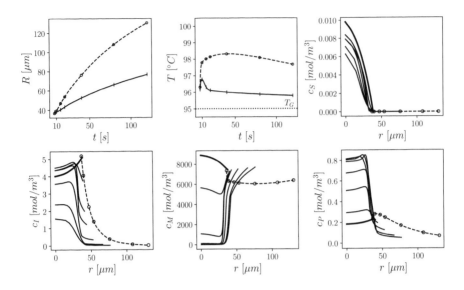

Figure 4.21: Droplet polymerisation with 90% monomer saturation in the drying gas at $10, 15, 20, 40, 80, 120$ s.

by monomer uptake - the concentration of living chains becomes lower and the monomer conversion rate is reduced. When the initiator is consumed completely (beyond the simulated time span), chemical reactions will break down and reactive absorption will stop. Even as the process is not finished after 120 s, the droplet radius is increased strongly as a result of monomer uptake. The excess temperature is elevated during the phase of reactive absorption, due to both the heat of absorption and chemical reactions. Polymer concentration is finally lower than at the beginning of this process phase as with decreasing initiator concentration longer chains are produced.

The distributed results exhibit very strong inhomogeneities in Figure 4.21. The principle course of the process at the drop's surface is similar to the lumped model, whereas the inhibition of transport evokes a completely different behaviour at the droplet's inner core. After less than 40 s process time full conversion is achieved there. In contrast, monomer is constantly absorbed from the drying gas at the droplet's outer rim so that its concentration remains on a nearly constant level. The low diffusivity of the polymer rich mixture prevents transport inside the drop so that chemical reactions of the absorbed monomer only

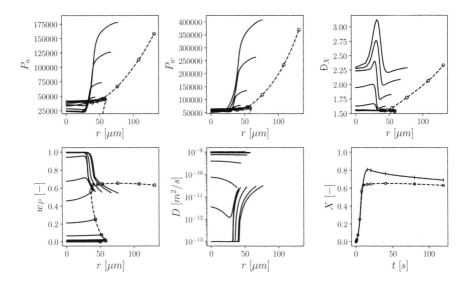

Figure 4.22: Radial profiles of polymer properties, diffusion coefficient and conversion for 90% monomer saturation in the drying gas at $2, 6, 10, 15, 20, 40, 80, 120\,\mathrm{s}$.

take place near the droplet's surface. With less monomer being consumed by reactions, the absorption rate is reduced likewise. Hence, the surficial monomer concentration is higher than in the lumped simulation. The effect of reactive absorption and the increase in droplet radius are therefore less pronounced in the distributed case and overestimated in the fully mixed model.

The polymer weight fractions and corresponding diffusion coefficients are depicted in the second row of Figure 4.22. In presence of monomer, the diffusivity at the droplet's surface stays at a comparably high level. Ongoing absorption leads to expansion of this outer volume. Transport within the inner core is inhibited with increasing polymer creation. Still, this nearly impermeable core grows slightly due to exchange of initiator and monomer near its transition to the monomer rich shell. The number and weight average of the chain length distribution exhibit a strong jump between the core with full conversion and the shell with absorbed monomer, as the monomer to initiator ratio is much higher and constantly increasing throughout the process at the droplet's outer part. Values of the lumped calculation are in the order of those near the droplet surface.

The long-time behaviour is depicted in Figure 4.23, as a hypothetical drop polymerisation process over half an hour at constant gas properties. Initial (0 s) and final profiles (1800 s) are drawn in bold. The progress of radius and conversion over time shows that reactive absorption does not stop before ten minutes process time - much longer than the typical residence time in a dryer. Surprising is the drop in conversion in the distributed calculation. This is caused by large gradients of the chain length distribution throughout the drop. As explained before, the high ratio of monomer to initiatior leads to the creation of very long chains at the droplets outer rim. The same number of polymer molecules therefore occupies a much larger volume near the surface and the total concentration is much than in the core. Fickian diffusion is driven by mole fraction gradients.

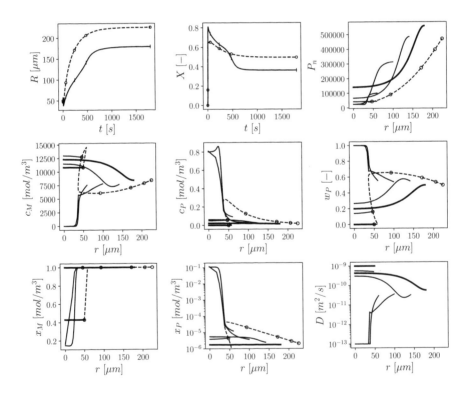

Figure 4.23: Long term evolution of the reactive absorption process for 90 % monomer saturation in the drying gas (t = 0, 4, 60, 240, 480, 1800 s, initial and final state in bold).

Flat mole fraction profiles involve a higher monomer concentration in the droplet's core than at its outer rim due to the different length of polymer molecules. Diffusion therefore leads to monomer transport from the drop's surface towards its centre (similar to the example in section 3.6.2). With the polymer in the core being diluted over time, the diffusion coefficient rises again. Diffusion is therefore accelerated and monomer transport further promoted. Ongoing polymer dilution while initiator is completely consumed leads to a drop in conversion. Finally, the polymer weight fraction is smallest at the core.

These profiles demonstrate the limits of a pseudo-binary Fickian diffusion concept. In the generalised driving force $\vec{d}_j = \sum_k \left(\delta_{jk} + x_j \left. \frac{\partial \ln \gamma_j}{\partial x_k} \right|_{T,p,\gamma_{l \neq k}} \right) \nabla x_k$ of the Maxwell-Stefan approach (equation 2.62), the activity coefficients in the mixture affect diffusive fluxes. The final profiles are therefore unlikely. Rather the monomer concentration will be a flat profile with values near the surficial one. Still, this effect mainly sets in on longer time scales. Profiles at 240 s still exhibit a polymer weight fraction near one at the droplet's centre. The principle diffusive behaviour of the process is hence modelled correctly for the relevant time scales of spray drying.

The basic course of the process is depicted in Figure 4.24 concerning the activities of solvent and monomer depending on weight fractions at the droplet surface ($P_n^{PAA} = 10000$). The process only takes place at the binary boundaries of the three-component system of solvent, monomer and polymer. In the first, rapid phase the binary solvent-monomer mixture is reduced to practically pure

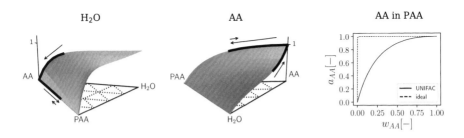

Figure 4.24: Activities of solvent and monomer using UNIFAC thermodynamics for $P_n = 10000$. Composition of the mixture throughout the process provided by bold line and arrows. Right graph: activity of acrylic acid in a AA-PAA mixture using UNIFAC and ideal thermodynamics.

monomer (+initiator). Polymerisation then leads to a monomer-polymer mixture. During the quasi-steady-state of reactive absorption, the surficial monomer content is lowered due to chemical reactions. In fact, it is the lowered activity at reduced monomer contents which causes the uptake of monomer from the gas phase. The smaller the monomer activity is, the smaller its partial pressure at the droplet surface is and the higher the driving force of monomer tranport from the surrounding gas to the droplet. After the initiator has been totally consumed, chemical reactions stop and the monomer content approaches an equilibrium at a final value which is elevated again.

While a high monomer partial pressure in the drying gas prevents monomer loss due to evaporation, full conversion is not achieved. Due to ongoing monomer uptake, a certain amount of monomer remains finally in the droplet, depending on its saturation in the drying gas. Moreover, the absorption period lasts for a very long time with respect to the residence time within a spray dryer.

Interim Conclusion

An elevated monomer content in the drying gas does not prevent monomer evaporation completely. Rather the process is divided into two stages, monomer evaporation and subsequent reactive absorption. Absorption kinetics have been taken into account reversely to evaporation and might need further investigation. In abundance of monomer in the gas, reactive absorption is an ongoing process.

4.3.6   Influence of Non-Ideality of Activities

The right graph of Figure 4.24 shows the activity of acrylic acid in a binary solution with PAA comparing UNIFAC calculations with ideal thermodynamics. With activities being identical to the mole fraction in the latter case, the AA activity stays at a level near one except for very low mass fractions, as the number of polymer molecules typically is orders of magnitude smaller than the amount of monomer molecules. Ideal thermodynamics therefore involve high surficial monomer pressures throughout the whole process except the very end, when the monomer mole fraction breaks down. As Figure 4.25 shows, this affects the model behaviour at high monomer saturations in the gas tremendously. Reactive absorption is taking place so scarcely in the distributed simulation that the droplet radius remains virtually constant after consumption of the monomer initially contained in the drop. Monomer uptake by absorption is again higher in the fully mixed droplet , yet to only a small degree compared to the previous simulations. In both the lumped and distributed calculations, the conversion approaches a value of practically one.

Number and mass average in Figure 4.26 show an inversed behaviour at the droplet surface compared to the previous calculation using UNIFAC. There, monomer uptake was so strong that very large molecules were created in the absorption phase due to the high monomer to initiator ratio. P_n and P_w profiles showed strong maxima at the drop's surface. In case of ideal thermodynamics, the uptake is scarce and the surficial monomer concentration small. Polymer molecules created by reactive absorption are very short, evoking a minimum in average chain length. Mole fractions profiles of monomer and polymer show strong gradients at the surface which are caused by absorption of monomer. Yet, this is only visible in mole numbers, whereas the polymer mass fraction is nearly one throughout the droplet. The diffusion coefficient stays at a low level everywhere and monomer infiltration into the droplet is prevented.

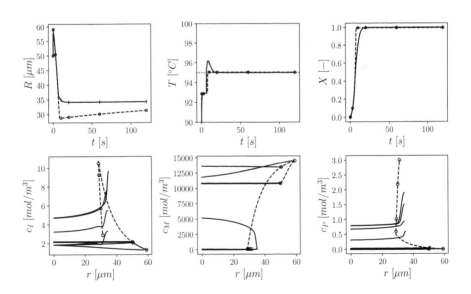

Figure 4.25: Course of the droplet polymerisation process assuming ideal mixture thermodynamics and a drying gas of 90 % monomer saturation ($t = 0, 0.3, 3, 10, 20, 60, 120$ s).

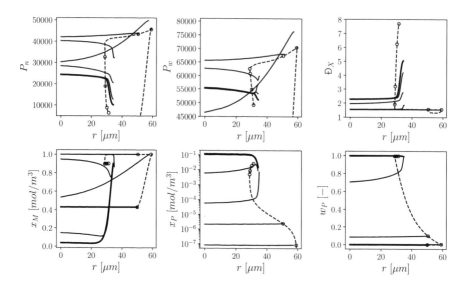

Figure 4.26: Polymer properties assuming ideal mixture thermodynamics and a drying gas of 90 % monomer saturation (t = 0, 0.3, 3, 10, 20, 60, 120 s).

Assumption of ideality of activities presumes that the occupied area of molecules at the drop's surface is proportional to their mole fraction. This is only valid for molecules of comparable size and not for solutions involving macromolecules. Whilst mixture thermodynamics of polymer solutions are clearly non-ideal and the example is rather of academic nature, it shows the impact of the thermodynamic model on the process behaviour of evaporation / absorption.

Interim Conclusion

Mixture thermodynamics play a dominant role for drying and reactive absorption. Simplification of the activity at the droplet's surface to an ideal behaviour can lead to drastically erroneous results. On the other hand, this also shows that reactive absorption in the previous calculation could have been less or more dominant, depending on the non-ideality of thermodynamics. Taking into account that ideal thermodynamics do not represent a polymer mixture well, the principle effect of reactive absorption can be taken for real.

4.3.7 Interaction with the drying gas

Gas properties were assumed as constant throughout all previous simulations. This is valid as long as the process is executed near the limit case of a single drop in an infinitely expanded volume, i.e. that the gas mass is very large compared to the droplet's one. Especially for the case of reactive absorption, mass and heat transfer between the drying gas and the droplet persist so that drying gas properties also may change throughout the process. This shall be approximated in the following by a simplified ansatz. If the total amount of mass of each volatile species and of heat being exchanged over the droplet's surface are integrated throughout the process and the initial drying gas properties (volume being in contact with the droplet $V^{G,0}$, partial pressures $p_j^{G,0}$ and temperature $T^{G,0}$) are provided, the current status of the drying gas can be obtained by algebraic equations:

$$p_j^G = p^G \frac{N_j^G}{\sum_k N_k^G} \tag{4.6}$$

$$T^G = T^{G,0} + \frac{Q^{lg}}{\sum_k N_k^G c_{p,k}^N} \tag{4.7}$$

$$N_j^G = N_j^{G,0} + \int_{t_0}^t 4\pi R(t) \Omega_j^N \, dt \tag{4.8}$$

$$Q^{lg} = \int_{t_0}^t 4\pi R(t) \dot{q}^{lg} \, dt \tag{4.9}$$

$$N_j^{G,0} = \frac{p_j^{G,0} V^{G,0}}{\Re T} \tag{4.10}$$

This set of equations considers the gas surrounding the droplet as a stirred tank reactor. A certain volume is attributed to each droplet, which can be considered as a function of the number of droplets per unit volume - a large droplet number corresponds with a low interacting gas volume and vice versa. This volume is considered as fully mixed. Moreover, gas volume and droplet are treated as a closed, adiabatic system. Evaporation and absorption of matter to/from the gas expands or shrinks the gas volume, which remains at constant pressure p^G. No heat is exchanged between the gas attributed to the drop and its surrounding. This model is of course an idealised simplification assuming all droplets within the dryer behaving in a uniform, equally distributed manner inside a gas exhibiting no gradients. Additionally, the relative motion between gas and droplets is

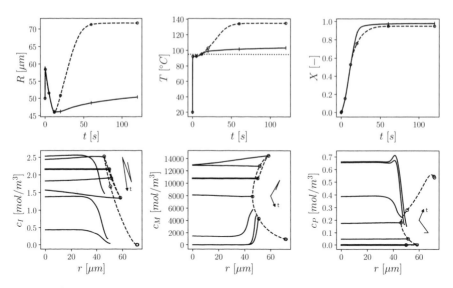

Figure 4.27: Droplet polymerisation with variable drying gas of 90% initial monomer saturation at $0, 0.5, 5, 12, 20, 60, 120\,\mathrm{s}$.

not taken into account. Yet, this simple model can be used to predict basically how strong the gas is affected by the droplet and how such droplet-induced changes feed back into the process of reactive drying.

Figure 4.27 shows calculations at the same process conditions as in Figures 4.20 and 4.21, but with variable drying gas of initial volume of 10^5 times the droplet volume at process start. Considering the gas density being three orders of magnitude lower than the liquids one, the gas mass amounts to about a hundred times the mass of the droplet. Similar to the previous simulation, after the initial solvent and monomer evaporation periods, reactive adsorption sets in and the increase of droplet radius is much higher in the case of full mixing than in the distributed model. However, compared to Figure 4.21, the effect is much less pronounced and ends early. The 0D calculation approaches a steady state after about 70 s with much less monomer uptake than in the case of constant drying gas. Monomer absorption is even far less visible in the distributed model. Despite the process has not reached an equilibrium after the calculated time of 120 s and monomer absorption will continue for a long time, changes of the radius and concentration profiles over time are very small, once, the monomer in the droplet's core has been consumed by polymerisation reactions. Moreover,

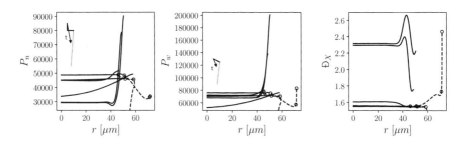

Figure 4.28: Polymer properties when the drying gas is variable at $0.5, 5, 12, 20, 60, 120\,\mathrm{s}$.

the conversion stays at a very high level near one in the distributed case after this point and approaches a steady-state value of 95 % in the lumped calculation. Generally, the drop's temperature rises throughout the time. As will be seen later, this is coupled with an increase of drying gas temperature. The increase is much stronger in the 0D simulation, as monomer absorption and reactions are much more pronounced there.

Profiles of number and weight averages and the dispersity of the chain length distribution are drawn in Figure 4.28. Similar to Figure 4.22 polymer chains are much longer at the droplet's surface than within its core. But, as only a minor amount of monomer is absorbed, this merely concerns a small shell, whereas in case of constant gas properties a very large outer volume is appended to the inner core. Additionally, the difference between shell and core is much smaller when the drying gas is variable.

Drying gas properties are depicted in Figure 4.29. Changes of solvent and monomer partial pressures are comparably small with respect to initial or saturation pressures of both components. Less than 1.5 and 7 % of the initial monomer content within the gas are transferred to the droplet in the distributed and lumped simulations, respectively. In contrast to the small alterations of solvent and monomer content, the gas temperature changes more strongly, especially in the lumped simulation. This is cause by monomer absorption and the heat of reactions. The amount of the different heat effects is depicted for the distributed calculation in the lower right graph. Droplet heat up naturally only plays a role at the very beginning. In the following period, cooling due to evaporation has the strongest impact. Both processes consume heat from the drying gas so that the accumulated energy transferred from the droplet to the gas is negative. However, the amount of heat created by polymerisation reactions rises strongly until monomer conversion in the droplet's core is finished after 25 s. Moreover, absorption of monomer additionally releases heat, yet to a smaller degree. The balance of all heat contributions becomes positve after 14 s. Chemical reactions play the major role concerning the total energy balance of droplet and drying gas. Differently to ordinary spray drying, the gas is in total not cooled because of evaporation but heated up due to the heat of reactions. Additionally, mass transfer takes place in both directions. Absorption processes partly recover the heat of evaporation. Even if the assumption of an adiabatic process is a simplification, the drying gas may leave the process at a higher temperature than within its feed, depending on process conditions.

As was discussed with respect to Figure 4.27, the monomer uptake is strongly slowed down when the drying gas is variable. The lumped model exhibits a steady state after about 70 s. Considering the drying gas properties in Figure 4.29, this is not caused by changes of the monomer content in the gas, which are rather slight. It is indeed the elevated temperature, which slows down absorption. The monomer saturation pressure depends non-linearly on the temperature.

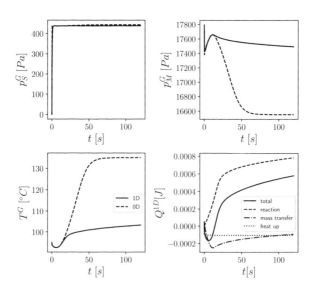

Figure 4.29: Properties of the variable drying gas and total heat transfered between droplet and gas (only distributed calculation).

Droplet and gas temperature are coupled. The temperature increase goes along with a strongly elevated monomer partial pressure at the droplet's surface. By this, the driving force for absorption is largely reduced. Surficial monomer concentrations are lowered and chemical reactions slowed down. Alteration of the drying gas by droplet processes hence moderates the effect of reactive absorption.

The proposed simple model of the gas as a stirred tank reactor strongly depends on the drying gas volume. The limit case of an infinite volume corresponds with the previous simulation of a constant gas, whereas the opposite example of the gas mass being similar to the droplet's mass involves exceedingly high changes. The ratio of gas volume with respect to the initial droplet volume was varied between 10^4 and 10^6 in Figure 4.30. Only distributed results are shown. A volume ratio of 10^6 is already near to the limit case of a constant drying gas. Then, gas properties in the second row change only slightly. The evolution of droplet radius, temperature and conversion is also very similar to the results in Figures 4.21 and 4.22 with constant gas properties. Monomer uptake is not stopped and conversion of just above 80% after 120s would not be

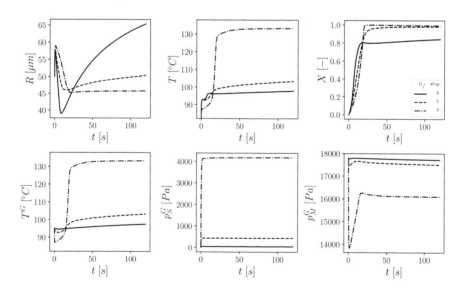

Figure 4.30: Effect of the gas volume interacting with a single droplet on the process and gas properties (distributed simulation)

applicable in a reactive spray drying process. If the volume ratio is low with a factor of just 10^4, the drying gas changes strongly. Other than before, changes of solvent and monomer partial pressures are much more pronounced due to the smaller gas volume. About 11 % of the initial monomer content in the drying gas is consumed throughout the process. The temperature rise is so high that, due to the high saturation pressure, reactive absorbtion practically stops early. The previously discussed case of a volume ratio of 10^5 lies in between. Concerning droplet radius and conversion it is comparable to the ratio of 10^4, with respect to droplet temperature and drying gas properties it is more similar to the 10^6 ratio. A constant or marginally changing drying gas will evoke the largest alterations due to absorption in the drop (leaving the droplet temperature out). On the other hand, the stronger the gas changes are, in particular concerning its temperature, the smaller the effect of reactive absorption is on the droplet.

Droplets in a real spray dryer are of different size and exposed to a non-uniform drying gas. The droplet number density will also change locally. The ratio of gas to droplet volume and gas properties hence varies within a spray dryer. Still, the principle cause-effect relationships that the gas changes coupled with processes inside the drop, especially reactive absorption, are covered well by this model. Moreover, these results provide a potential technical solution to the problem of ongoing monomer absorption. As the temperature controls monomer absorption, the droplets need to be heated up shortly before the end of the process. The surficial monomer pressure will rise and absorption will stop. The remaining amount of monomer will be converted to polymer at high temperatures very quickly. Technically, it is not trivial to provide a rapid and preferably uniform heating near the end of the process, but a smart solution which efficiently brings absorption to an end and affects final product properties only to a minor measure. Possible technical implementations might involve blowing in of hot drying gas or microwave heating devices for droplet heat up.

Interim Conclusion

The effect of the drying gas has been modelled in an approximate manner. This example shows the impact of the process on the gas, which may alter conditions such that reactive absorption ends. Moreover, the model reveals a difference of reactive drying to conventional spray drying. The energy balance is substantially different due to the heat of reaction. Increase of the gas temperature stops absorption processes and is therefore an additional lever for process control.

4.3.8 Applicability of the QSSA model

All simulations so far have been performed using the method of moments. The Quasi-Steady-State Assumption model presented in section 3.3.3 is simpler, as the polymer is only represented by two variables instead of six moments - the concentration of dead chains c_D and of the monomer units being incorporated in these chains $c_{D,M}$, identical to the zeroth and first moment of the dead chains' distribution. Living chains' properties and polymerisation reactions are provided as algebraic expressions depending on initiator and monomer concentrations. The QSSA hence introduces less interactions between variables in simulations. Calculations are therefore significantly faster. On the downside, the QSSA is not as versatile as the method of moments with respect to chemical reactions. Moreover, only the number average of the chain length distribution is calculated, whereas its weight average and dispersity remain unknown. For a simple reaction scheme as it is employed in this work, results of the QSSA and the method of moments should behave identical otherwise.

Figure 4.31 shows a QSSA simulation with respect to the previous test case of polymerisation involving monomer evaporation into a gas with high monomer

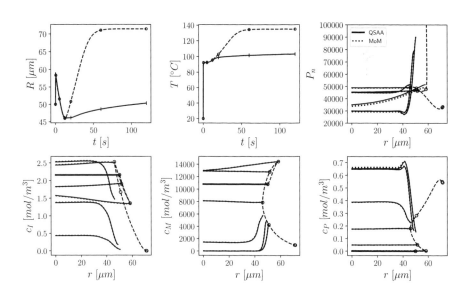

Figure 4.31: Course of the process employing the QSSA, conditions identical to Figure 4.27.

saturation and consideration of gas alterations throughout the process. This is the most complicated example with the highest number of physical effects and therefore a good test case for the applicability of the QSSA. Calculation of activities were performed according to UNIFAC. The dotted profiles show corresponding results obtained by the method of moments. Both models exhibit an equivalent behaviour. Profiles of initiator, monomer and polymer concentrations throughout the process are practically identical. Likewise, the calculated number averages of the chain length distributions match very well with the exception of visible discrepancies of the first profiles at 0.5 s. The assumption of a quasi-steady-state is not strictly valid at this early instant of time, because the process undergoes large changes during the first second. Nevertheless, this only affects a short period throughout the process and a vanishingly small amount of polymer. Computation times were 2754 s using the method of moments and 930 s for the QSSA model on a third generation Core i5 mobile CPU running at 2.5 GHz (60 nodes in radial direction).

Interim Conclusion

Provided that polymerisation reactions are simple, the Quasi-Steady-State Assumption provides an approximation, which is as good as the method of moments. The latter is more versatile concerning polymer properties and chemical reactions. Hence, the QSSA is a good alternative as long the reaction scheme is not complicated and mainly the number average of the chain length distribution is of interest. In other cases it may still serve as a screening method before employing detailed simulations.

4.4 Summary of Basic Findings on Droplet Polymerisation

The residence time within a spray dryer is very short. Spray polymerisation therefore demands very fast chemical reactions. Due to the cooling effect of drying, evaporation of a solvent reduces the droplet temperature and hinders chemical reactions. Before polymer is built, the largest part of the solvent has vanished, if the drying gas exhibits a low solvent saturation. Polymerisation is therefore carried out in bulk. The assumption of concurrent solvent drying end chemical reactions is not confirmed by the simulations in this work. The process rather takes place in subsequent steps of drying and chemical reactions afterwards. Kinetic data for reaction conditions in the droplet may be hard to obtain and require extrapolation of literature data.

As long as the monomer does not evaporate, bulk polymerisation in the drop takes place in a very uniform way and is described sufficiently by a lumped model considering the droplet as fully mixed. Monomer vaporisation introduces spatial inhomogeneities. In this case, evaporation of monomer and chemical reactions in fact take place at the same time, with the share of both processes depending on the monomer saturation in the drying gas. This case is only appropriately described by a distributed model. If the monomer saturation in the drying gas is low, the largest part of the initial monomer is lost to the drying gas. Moreover, as the initiator concentration rises tremendously, very many short chains are created and a product of very high dispersity is obtained. Even with potential monomer recovery, this appears unfavourable.

The monomer loss to the drying gas can be limited by pre-polymerisation. If already a certain amount of polymer is created before atomisation, drying will leave a polymer rich outer shell at the droplet's surface with a low diffusivity. Evaporation becomes hindered so that a larger amount of monomer can be converted to polymer. It is sensible to further investigate this process variant mainly for bulk polymerisation, as not only monomer vaporisation is prevented, but also solvent drying.

Another option for limiting monomer loss involves a high monomer saturation in the drying gas. In doing so, the driving force for mass transfer is limited so that polymerisation reactions are in favour over evaporation. With the surficial monomer fraction being reduced by chemical reactions and evaporation, its partial pressure at the droplet surface is also lowered und undershots the monomer pressure in the surrounding at some time so that mass transfer is inverted. In the following phase, monomer is absorbed from the gas and converted into polymer. This process of reactive absorption is strongly depending on mixture thermodynamics at the droplet's surface. In the - unrealistic - case of an ideal mixture, monomer absorption does not play a role. Moreover, interactions between drying gas and the droplet strongly affect the duration of this final process period and the amount of monomer uptake. In the limit case of a constant drying gas, theoretically very large amount of monomer would be absorbed and the process would take much longer than the typical residence time in a dryer. On the other hand, chemical reactions heat up droplet and drying gas. Due to the thereby increased monomer partial pressure at the droplet's surface, absorption is hindered and the monomer uptake limited. Reactive absorption can therefore be stopped if the droplet temperature is increased. Whilst absorbed monomer is for the largest part converted into polymer as well, this reaction takes place, when the initiator content is already low, due to both prior decomposition and dilution by monomer. The chain length distribution therefore exhibits inhomogeneities between the droplet core not being affected by absorption and the outer shell.

An interesting idea is to run the drying gas in circulation. Its monomer saturation would self-adapt the the spray conditions so that over some cycles the system will reach a steady-state, in which monomer absorption and evaporation during different process phases are at an equilibrium throughout the whole process. All monomer provided initially in the feed will be converted to polymer on average of all droplets. However, this is only feasable to systems of bulk polymerisation as otherwise the solvent will accumulate in the drying gas.

As long as the reaction scheme is as simple as in the present simulations and if only P_n values are of interest, the Quasi-Steady-State-Assumption sufficiently describes the system. QSSA calculations are significantly faster. In more complicated cases QSSA models can still be used for screening of the process. Lumped models are only applicable, when drying and diffusional transport limitations do not interfere. This is the case when monomer evaporation does not play a role. In other cases, only a distributed simulation reveals all details of the process and lumped calculations can exhibit large errors. 0D results often only match surficial 1D values, with the distributed model predicting very different results inside the droplet.

4.5 Process Evaluation, Numerical DoEs

The previous examples provide insight in the basic mechanisms during single dropled polymerisation in a spray dryer. Yet, the question how the process depends on various parameters remained open. Design of Experiment (DoE) is a tool in order to evaluate correlations between product properties and process parameters. Being originally applied to measurements, it is a valid method in order to examine the behaviour of a numerical model with respect to several parameters. The advantage of this approach is that model responses can be revealed, which are not visible in one-factor-at-a-time variations. In all parameter studies, the initiator to monomer ratio I/M, gas temperature T^G and its saturation with monomer Ψ_M^G, the initial droplet radius R^0, the ratio between gas and droplet volume V^G/V^d and the diffusion parameters w_P^{crit} and $D_{log_{10}}^{lim}$ have been varied. Three different numerical DoEs will be evaluated in the following:

- Polymerisation in solvent: Additionally, the initial monomer mass fraction w_M and the solvent saturation in the drying gas Ψ_S^G were varied.

- Bulk polymerisation: As solvent evaporates early, the model behaviour in complete abscence of solvent is evaluated. (No additional variables were varied.)

- Pre-polymerisation: The impact of polymerisation prior to atomatisation is studied by the extra parameters temperature $T_{prepoly}$ and time $t_{prepoly}$.

4.5.1 DoEs' Setup and Evaluation

Distribution of DoE Points

When used in conjunction with measurements, DoE approaches are often set such that the number of expensive experiments is minimised whilst the desired interrelations can still be obtained. The kinds of correlations (linear, quadratic, parameter interactions), which the DoE shall be able to predict, need to be chosen beforehand and determine the parameter settings used in the experiments. Boundary points are often favoured in order to cover a large range of parameters (e.g. D-optimal designs) without the need of extrapolation.

When numerical simulations can be carried out with comparably small computation times, such limitation are irrelevant and a broad variety of parameter

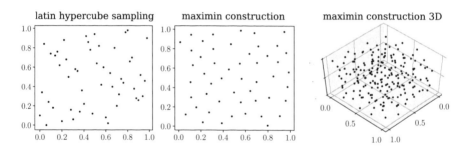

Figure 4.32: Latin hypercube sampling vs. maximin construction in 2D, 3D maximin example

settings can be applied. This is particularly useful, if strong non-linearities and interactions between parameters need to be evaluated. The settings are then chosen such that the whole parameter space is covered sufficiently and fairly regularly. In the following studies, points were set according to a maximin construction taken from the Python diversipy package, described in Wessing (2015). This algorithm maximises the minimum distance between sample points and ensures a better coverage of the parameter space than typical latin hypercube sampling. The sampling of single variables may be irregular, though, with some value ranges being chosen more often than others. An example is provided in Figure 4.32. The left frame shows latin hypercube sampling of 50 points in two dimensions, the middle one a corresponding maximin construction. The right frame provides a maximin construction in three dimensions by 200 points.

The effects of process/input parameters on the outcome were modelled via multiple regression using Gaussian Processes. The theory is shortly described in appendix A. Further insight is provided by Rasmussen and Williams (2006). All calculations were carried out with the lumped and the one-dimensional model (employing 20 finite volume cells of equal volume).

Characteristic Values for DoE Evaluation

The simulation data needs to be condensed into characteristic values before a regression can be undertaken. Such values are the conversion X - with its reference value continually changing due to monomer evaporation and absorption - and the yield Y - the amount of monomer being converted into polymer compared to the monomer initially contained in the droplet:

$$X = \frac{m_P}{m_M + m_P} \qquad (4.11)$$

$$Y = \frac{m_P}{m_M (t = 0)}. \qquad (4.12)$$

Absorption and condensation of monomer may cause the final polymer mass to be higher than the initial monomer mass. In this cases, the yield rises above 100 % with respect to the initial monomer content in the droplet, as the drying gas acts as an additional feed. Other values are polymer properties and polymer mass per droplet. When solvent is present at atomisation, the product's moisture content is also relevant:

$$\Psi_S = \frac{m_S}{m_P + m_M}. \qquad (4.13)$$

A rough estimation of the travel distance / falling height covered during the process can be provided in simulations' post-processing by piecewise summation of the distances traveled between each instant of time using the terminal velocity according to Stokes' flow (equation 2.118)

$$h_{fall}(t_i) = \sum_{j=1}^{i} \frac{2g}{9\eta^G} \left(\rho^L - \rho^G \right) \bar{R}^2_{j-1,j} \left(t_j - t_{j-1} \right). \qquad (4.14)$$

This is a very coarse value, as it generally neglects the gas motion around the droplet and the three-dimensional flow field in the dryer with effects like recirculation, but still provides a qualitative information about the residence time and dimensions required by the combination of process parameters.

Regression of such values can for instance be caried out with respect to certain instants of process time (e.g. conversion after 10 s). However, the residence time in the dryer strongly depends on the droplet size, which for its part is affected by process conditions. Figure 4.33 shows an example regression of the height of fall after 10 s in droplet polymerisation within a solution. The vertical dashed lines represent the point of prediction (50 % monomer mass fraction in the feed, an initiator to monomer ratio of 10^{-4} etc.), the solid lines the predicted value when varying one parameter and the dotted ones the 95 % bounds of the Gaussian process in the fit. The fall distance - inverse to the residence time - may differ strongly, not only depending on the initial radius, but also on the saturation of solvent and monomer in the drying gas. It is therefore not reasonable to compare parameter variations at a certain instant of time, when in one case

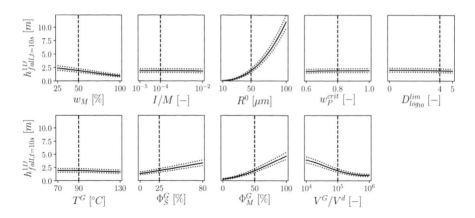

Figure 4.33: Example evaluation of a drop's falling distance at polymerisation in solvent after 10 s process time.

the droplet already left the dryer and in another case this represents just a fraction of the total process time. Evaluation of characteristic values with respect to a certain falling height provides therefore more meaningful results concerning the applicability of process parameters.

As the numerical model of droplet polymerisation exhibits the aforementioned uncertainties, only the principle behaviour of the process can be depicted. Assessing values with respect to a distinct falling height means evaluation at a comparable position within the dryer, not distinctly at the respective height. The numerical DoEs reveal principle, qualitative features concerning the applicability of droplet polymerisation within a spray.

4.5.2 Droplet Polymerisation with Solvent in the Feed

As polymerisation within a spray under bulk conditions involves less additional steps (like separation of solvent und vaporised monomer in the drying gas), introduction of a solvent appears only sensible in case of requirements such as dissolution of salts like NaA. Numericals experiments in sections 4.2 and 4.3 showed a rapid decrease of solvent content due to drying at the very beginning of the process. A similar observation was made by Franke, Moritz, and Pauer (2017), who observed precipitation of low-molecular NaA salt.

Requirements for polymerisation within a solvent are therefore:

- Process conditions are significantly different to bulk polymerisation. If salts are present, the solvent stays at a sufficiently high level to prevent precipitation during the reaction period.

- The remaining moisture content at the end of the process is low and expensive further drying is prevented.

It is obvious that these are conflicting goals.

Figure 4.34 shows a predicted plot of the droplet polymerisation in presence of a solvent for characteristic values after 5 m. Additionally to the yield and conversion, the average solvent concentration during polymer creation and the moisture content are shown.

The yield of the process is strongly improved at increased initiator to monomer ratios, small droplet radii and a higher monomer partial pressure in the drying gas. All three parameters are linked to reactive absorption of monomer from the drying gas. Increased consumption of monomer in chemical reactions due to an increase of chain radicals will intensify the rate of absorption of monomer from the gas phase. Small droplets have a much longer residence time so that the monomer take up is prolonged. A higher monomer partial pressure in the gas increases the driving force for absorption. Whereas the gas temperature also has a strong effect on the yield, it is limited insofar that boiling needs to be prevented. The current model does not consider boiling directly, which needs to be evaluated within post-processing.

The conversion after 5 m is for one thing influenced by typical process parameters from polymer reaction engineering as the initiator to monomer ratio and the temperature, for another thing by parameters affecting the droplet radius and therefore the residence time. Besides the initial radius these are in particular the saturation of monomer and solvent in the drying gas, which control the evaporation and absorption rates of both components.

The moisture content Ψ_S is mainly affected by the solvent content in the drying gas, except for very low monomer saturations, which favour monomer evaporation over solvent vaporisation. At the prediction point of 25 % solvent saturation in the gas, the moisture content at 5 m falling distance is fairly low. At the same time, the average solvent concentration in the droplet is also greatly diminished compared to its initial value. $\bar{c}_S$ is a weighted average with respect

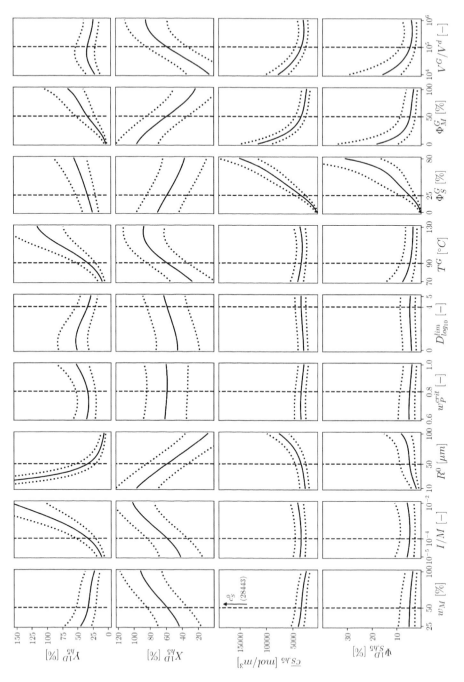

Figure 4.34: Evaluation of polymerisation in solvent at 5 m height of fall.

to the polymer creation at evaluation points i during post-processing

$$\bar{c}_S = \frac{\sum_j^i \left(m_{P,j} - m_{P,j-1}\right) \frac{c_{S,i-1}+c_{S,i}}{2}}{m_{P,i}}. \tag{4.15}$$

In doing so, process periods with sole drying or in which reactions have stopped are not taken into account. For the parameter combination in the plot, the predicted average solvent concentration $3400\,\mathrm{mol/m^3}$ is only a small fraction of the initial solvent concentration of $14500\,\mathrm{mol/m^3}$. This corrsponds with a virtual monomer mass fraction of about $94\,\%$ so that the process is nearly carried out under bulk conditions. This shows the principle problem of spray polymerisation in presence of a solvent: Drying needs to be strongly inhibited, if polymerisation shall not be carried out in bulk. Possible solutions in order to provide a sufficiently high solvent concentration are:

- Continuous increase of the drying gas temperature throughout the process: Process conditions change from mild to harsh drying conditions so that the solvent content in the droplet is high at the beginning, when precipitation is to be hindered, decreased, when the solute content is already limited, and very low at the end, when a dry product is required. In order to prevent bursting of the droplet, boiling of the remaining solvent has to be avoided. As evaporation rates are small after the initial rapid drying phase, droplet and gas temperatures are nearly identical. Possible temperature profiles in the dryer are therefore limited by the boiling point of the mixture.

- If only a small solvent concentration is required during chemical reactions, this can be achieved by adjusting the solvent saturation in the gas like at the prediction point in Figure 4.34. The remaining moisture content is fairly low then. Yet, process design needs to account for inhomogeneities within the drying gas in order to prevent locally harsh drying conditions.

In any case, process conditions are such that the drying gas needs to be preconditioned in order to provide certain solvent and monomer saturations. Afterwards recovery of monomer from the gas is necessary. Even taking all model uncertainties into account, droplet polymerisation in presence of a solvent remains a very limited process variant due the target conflict of (partly) preventing solvent evaporation first and promoting it afterwards, when a polymer hull around the droplet decreases the solvent permeability. This conflict may be resolved as long as the required solvent concentrations are low, but the possible process window is very narrow.

4.5.3 Bulk Polymerisation within a Droplet

With control of the solvent concentration being a challenge, bulk polymerisation appears a more favourable process option. An overview on process characteristics for polymerisation in bulk is provided in Figure 4.35. The yield is mainly affected by drying and absorption. Just as in the previous example of polymerisation in solvent, monomer uptake is increased due to longer residence times of small droplets and increased monomer saturations in the gas. The diffusion parameter $D_{log_{10}}^{lim}$ controls the maximum order of magnitude at which the diffusion coefficient is decreased (at the point of prediction from 10^{-9} to 10^{-13} m^2/s). The higher the permeability of the polymer hull is, the more intense the transport of absorbed monomer towards the droplet's core will be and the higher the conversion rate in reactive absorbtion. section 4.3.5 already showed that the monomer uptake is much higher in the 0D approximation (which corresponds with $D_{log_{10}}^{lim} = 0$ in the 1D model). Knowledge about the evolution of diffusivity with ongoing polymerisation is therefore crucial for an accurate prediction of the process outcome especially when absorption from the gas may play a role.

The conversion after 5 m falling distance is generally rather high. In regimes of ongoing monomer absorption it is decreased as the reference monomer amount for conversion calculation increases continuously. The maxima in the conversion curves concerning the initiator to monomer ratio, the droplet radius and the gas temperature are caused by a runaway of the chemical reactions when these values are increased further. This is indicated by the maximum temperature in the process, which rises to very high values in these cases. As has been stated before, the boiling point of the mixture was not considered in the simulation which explains values beyond the boiling temperature of pure AA. However, the regression runs short in predicting the runaway behaviour in detail as it is a switching-type effect when increasing process parameters.

The runaway behaviour is shown in 4.36 by a one-factor-at-a-time variation of the three above-mentioned process parameters in the left graph. The right graph exhibits the course of initiator concentration, droplet temperature and radius over time for the point depicted in the initiator to monomer variation on the very left. Other parameters had been set to 85 % monomer saturation, $w_P^{crit} = 0.8$, $D_{log_{10}}^{lim} = 4$ and a ratio between gas and droplet volume of 10^5. The course of the process over time reveals the reason for the runaway behaviour. The early droplet heat-up period with slight monomer condensation is followed

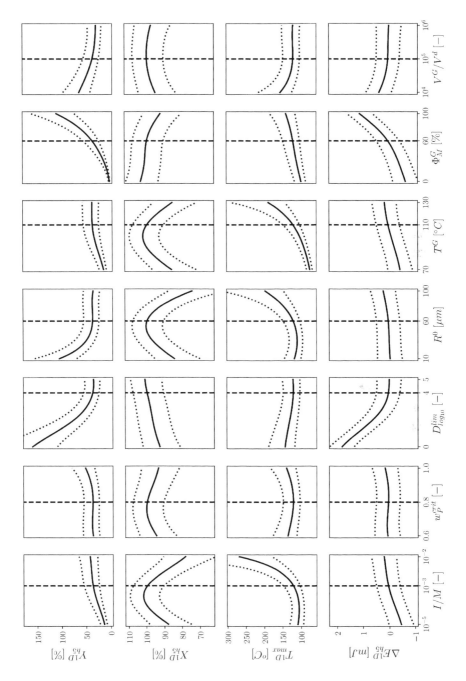

Figure 4.35: Process evaluation for bulk polymerisation at 5 m height of fall.

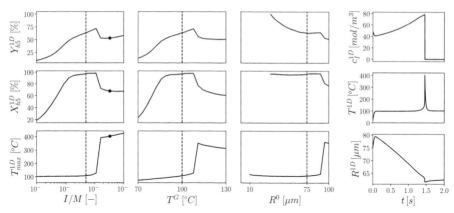

Figure 4.36: Runaway behaviour of the polymerisation process: yield, conversion and maximum temperature in a one-factor-at-a-time variation (left graphs), course of the process over time (right) for the depicted point on the very left.

by monomer evaporation and at the same time polymerisation reactions. The heat of reaction can be consumed by monomer vaporisation - an increase in droplet temperature will go along with enlarged mass transfer and evaporative cooling. With enough polymer being created, transport of monomer to the droplet surface is hindered and evaporation decreases. With less cooling due to vaporisation, the droplet temperature rises, chemical reactions are accelerated and the heat of reaction cannot be transferred completely to the surrounding gas anymore. Consequently the reactions run away until all initiator is consumed. This effect is remarkable as heat exchange of a small droplet with its surrounding is very intense due to the high surface area to volume ratio. Just for that reason, the droplet returns to the temperature of the surrounding gas within a fraction of a second after the temperature peak. Still, the heat release by chemical reactions can be uncontrollably high under certain process conditions.

Taking the boiling temperature of the mixture into account, the runaway behaviour will be superimposed by boiling and possibly bursting of the droplet so that the maximum temperature will be limited. Still, the outcome that the process runs out of control is the same and corresponding parameter settings need to be avoided. This affects in particular cases in which the reaction rate is very high and mass transfer to the droplet's surface can be strongly limited - a large initiator content and high temperatures or a low diffusivity of the polymer hull and a large droplet radius. The runaway is not captured by a lumped model, as either the surficial monomer content is high enough due to (falsely) full mixing to provide sufficient evaporative cooling or the monomer content is too small to maintain a reaction being out of control. If the diffusion coefficient stays at a sufficiently high level, this challenge will therefore vanish.

The heat release by chemical reactions is significant. The fourth row in Figure 4.35 shows the energy being set free per droplet. The heat of the polymerisation reaction (77.5 kJ/mol) is higher than the heat of evaporation (47.5 kJ/mol) so that at slightly less than 40 % yield and 100 % conversion the droplet's energy balance throughout the whole process sums up to zero. This is different to typical spray drying, which is energy-intensive like all drying operations. This peculiarity provides some flexibility concerning the yield per droplet. Two process variants of bulk polymerisation appear favourable:

- Run the process at a yield which requires low additional energy input and recover the evaporated monomer from the drying gas:
  If the monomer saturation after recovery equals its value at the dryer's inlet, the drying gas can be recirculated. From an energetic point of view heat losses to the surrounding and accompanying processes like gas preconditioning need to be compensated. The processes within the total gas and droplets will run energy-self-sufficient at yield values of about 40 % or higher.

- Run the drying gas in recirculation without monomer recovery:
  The process will adjust itself automatically to a drying gas saturation at which the yield is 100 %. As long as the yield is smaller, monomer will accumulate in the drying gas until absorption and evaporation are in balance, for a monomer content too high it will be the other way round.

In both cases reactive absorption should be prevented at the end of the process in order to achieve a product of 100 % conversion at the dryer's outlet as discussed earlier. The second solution is simple and elegant, as no monomer is lost, no additional steps for recovery and preconditioning of the gas are required and the process is intrinsically controlled. Yet, the challenge relies in the very high monomer saturation of the gas. For one thing, condensation of monomer at cold spots has to be prevented, for another thing thermal polymerisation in the gas phase due to impurities might be an issue.

The monomer saturation in the drying gas in order to achieve 100 % yield after a falling distance of 5 m is depicted in Figure 4.37 depending on gas temperature and droplet radius. The contour lines of the isosurface in the $T^G - R^0$ plane refer to monomer saturations in the gas between 45 and 100 %. Other parameters have been kept at an I/M ratio of 10^{-4}, $w_P^{crit} = 0.8$, $D_{log_{10}}^{lim} = 4$ and a V^G/V^d ratio of 10^5. This graph is speculative in terms of absolute numbers and only predicts the qualitative features of the process. It is obvious that the droplet radius, at which 100 % yield can be achieved, is limited if a certain monomer saturation may not be exceeded. Moreover, a collective of droplets of various size will experience a different individual yield per droplet. Generally, all drops surpass a period of evaporation. A high yield near 100 % will only be achieved by subsequent absorption. Smaller drops take up more monomer due to the longer residence time and exhibit a yield even beyond 100 % (their parameter combinations lie above the isosurface), larger droplets absorb less monomer and exhibit a net loss of monomer. As a positive result, the droplet size distribution could therefore narrow throughout the process. On the downside, even if the average yield is at 100 %, a distinct fraction of the droplets might leave the dryer prematurely. 100 % conversion thus need to be ensured concerning all relevant droplet sizes.

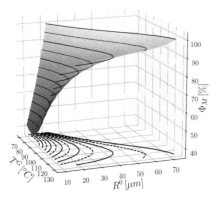

Figure 4.37: Isosurface and contour lines for 100 % yield in bulk polymerisation depending on droplet size, gas temperature and monomer saturation.

Figure 4.38 exhibits the differences between the distributed and the lumped model by the ratio of the respective yield $Y^{1D/0D} = Y^{1D}/Y^{0D}$. Generally, both models differ strongly as long as diffusion is limited in consequence of polymer creation. All parameters exhibit a strong interaction with the maximum limitation of the diffusion coefficient so that for $D_{log_{10}}^{lim} = 0$ most of the differences between both models vanish. The 0D model thus falls short of predicting the process behaviour appropriately and 1D simulations are commonly advisable.

Finally, the predicted polymer properties shall be shortly discussed in order to highlight the model's capability to simulate characteristic values of the polymer and local inhomogeneities. Predictions of number and mass average and dispersity are plotted in Figure 4.39. Additionally the droplet is split into an outer

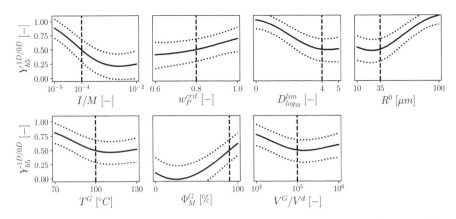

Figure 4.38: Comparison of 1D and 0D simulation by ratio of polymer yield.

shell and an inner core of identical volume and the ratios between shell and core for P_n and P_m are depicted in the lower rows. Values again refer to 5 m falling distance. The prediction point is chosen such that 100 % yield are achieved. The dependence of number and mass average on the initiator content is just as expected. The fewer chains are initiated, the longer these chains will grow. This applies similarly to low gas temperatures as kinetics of chain initiation display a higher temperature dependence than propagation reactions.

Parameter settings with a tendency to runaway ($I/M \uparrow$, $R^0 \downarrow$, $T^G \uparrow$) exhibit a decrease in number average. During runaway a very high number of small chains is created which decrease the average chain length, but scarcely contribute to a mass-weighted average. This is also visible in the dispersity, which rises in case of all undesirable process condition - besides the runaway settings a small monomer saturation in the gas, as discussed in section 4.3.3. Comparing number and mass average in the droplet's shell and the core, chains are generally longer at the outer rim. This is due to reactive absorption, which provides an ongoing influx of monomer molecules at a decreasing initiator fraction in the vicinity of the droplet's surface. The chains being created are therefore continuously getting larger and the ratio between shell and core chainlength is strongly dependent on the monomer saturation. The maximum near 50 % saturation implies that the complete outer half of the final droplet is created by reactive absorption. If reactive absorption is even stronger, the largest part of the product comes from this process period so that in total it is more homogeneous again.

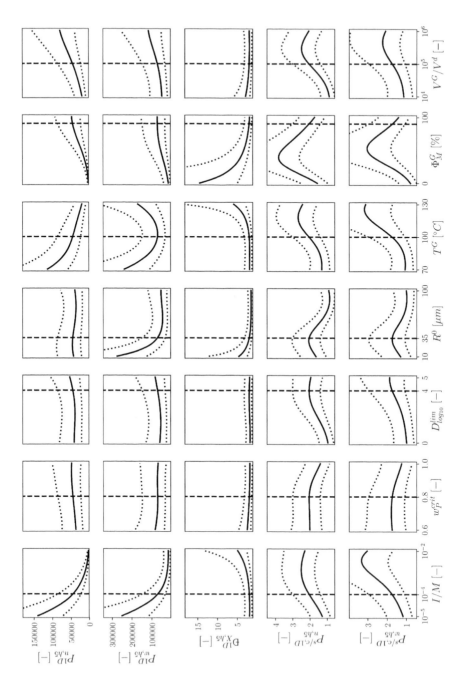

Figure 4.39: Characteristic values of the chain length distribution in bulk polymerisation.

4.5.4 Bulk Feed with Pre-Polymerisation before Atomisation

Pre-polymerisation in presence of a solvent does not appear as a favourable process variant due to hindered solvent evaporation (see section 4.3.4). In the following, the case of pre-polymerisation under bulk conditions shall be further evaluated. Therefore, the time and temperature during polymerisation before atomisation have been varied additionally to the parameters used before ($0\,\text{s} \leq t_{prepoly} \leq 7\,\text{s}$, $70\,°\text{C} \leq T_{prepoly} \leq 130\,°\text{C}$ in order to be in a similar range as for T^{G}). Both parameters determine the initial polymer conversion at droplet formation $X_{prepoly}$. This value was used for the regression of process features, as the effect of a certain amount of polymer being present at droplet creation on the process is clearer to understand than time and temperature of an upstream process.

An overview of the process is provided in Figure 4.40. Pre-polymerisation indeed acts as a lever to increase the yield. The relative gain per additional polymer created before atomisation is largest for small and moderate conversions $X_{prepoly}$. Interestingly, the yield at the point of prediction is not exceedingly high, but achieved without additional monomer in the gas phase. In pure bulk polymerisation ($X_{prepoly} = 0$), the yield would be about one percent at the same process conditions. Whereas pre-polymerisation directly provides 20 % of yield (equal to $X_{prepoly}$), around the same amount of additional yield is achieved by alteration of transport conditions within the drop in providing polymer right at the start. It is clear that this effect is most pronounced when diffusion is strongly hindered by polymer creation ($D_{log_{10}}^{lim}$ ↑) and vanishes in ideally mixed drops (see also section 4.3.4).

The other values in Figure 4.40 can be explained briefly. The conversion is generally at a high level as - besides for large droplets or large monomer saturations - the residence time is long enough that the monomer is fully consumed by evaporation and chemical reactions. The temperature and energy balance fits scatter significantly so that only clear features can be regarded. These are temperature maxima in regions which are prone to runaway and a minimum in case of no prepolymerisation (and hence nearly full monomer evaporation). Correspondingly, the overall energy balance exhibits the same minimum and is otherwise nearly balanced. At high amounts of pre-polymerisation, chemical reactions are shifted prior to the spray process and the energy release in the drop remains unaffected.

The spatial inhomogeneity of the polymers number average is depicted in

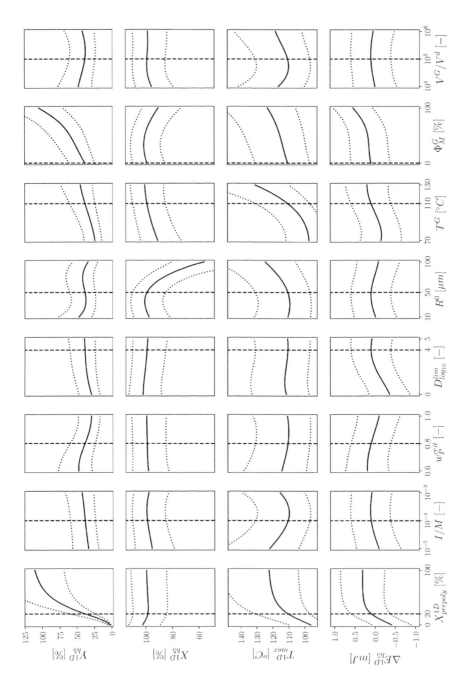

Figure 4.40: Process evaluation for partly pre-polymerisation in bulk.

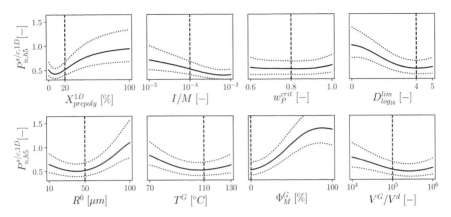

Figure 4.41: Inhomogeneity of the number average between droplet shell and core when pre-polymerisation is applied.

Figure 4.41 at the same point of prediction. Remarkably, the ratio between shell and core is inverted to the prior simulation of pure bulk polymerisation. This is due to the fact that reactive absorption does not take place as the monomer saturation in the gas was set to zero. Polymerisation reactions therefore occur simultaneously with monomer evaporation, by which the initiator content at the surface is elevated. Hence, smaller chains are created at the droplet's surface than in the shell. The behaviour changes for high monomer saturations, which evoke absorption of monomer with the differences between shell and core as described in the previous section.

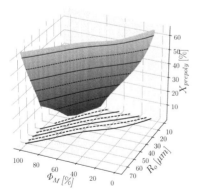

Figure 4.42: Isosurface of 100% yield via pre-polymerisation.

When 100% yield are desired (e.g. by recirculation of the drying gas as described above), prior polymerisation may widen the process window. Figure 4.42 again shows an isosurface of 100% yield depending on droplet radius, saturation with monomer and the conversion after pre-polymerisation. The contour lines in the $R_0 - \Phi_M$ plane refer to $X_{prepoly} = 10$ to 60%. Other process parameters are $I/M = 10^{-4}$, $T^G = 110\,°C$, $w_P^{crit} = 0.8$ and $D_{log_{10}}^{lim} = 4$. The maximal droplet radius at a certain monomer saturation in

151

the drying gas is raised by 10 to 20 μm for 20 % of conversion before atomisation. For large $X_{prepoly}$ values the gain is higher, but these appear of rather academic nature because atomisation of a strongly polymer loaden solution brings its own challenges along.

Summing these results up, pre-polymerisation may be useful in two aspects:

- Droplet polymerisation within a gas containing no monomer:
  Slight polymerisation may raise the yield in unloaden gas strongly, if the diffusion coefficient within the polymer is sufficiently low. Challenges arising from an elevated monomer content in the gas are avoided. Preconditioning of the gas and recovery of monomer can be implemented in a simpler way. Due to the heat of reaction, the energy demand within the dryer is low or even energy is released, depending on the final yield.

- Enhancement of the process window at higher monomer contents within the gas:
  Pre-polymerisation may either be used to increase the yield at otherwise constant process conditions or enlarge the applicable range of process parameters. The relative gain is yet highest in cases of very small yield if no pre-polymerisation was applied.

In any case, polymerisation prior to atomisation is an additional process step, which itself needs to be designed carefully and involves further invest. Moreover, if a pre-polymerisation step is installed anyway and well-controlled, the question is whether polymerisation in a spray is really advantageous over spray drying a diluted polymer solution, which has been processed up to full conversion beforehand.

4.6   Discussion and Suggestions for Further Research

The simulations show that the presented distributed model is capable of predicting polymerisation within a drop sufficiently. Current challenges are the uncertainty of the data the model is operating with. Whereas the model analysis has been performed to the author's best knowledge and the qualitative features should be predicted correctly by the present model, the process peculiarities - like the region in which reactive absorption takes place - might be shifted over the parameter range. Experimental validation of the proposed process behaviour is therefore necessary. This means in particular:

- Solvent concentration within the drop during polymerisation:
  Numerical results show clearly that the solvent content during polymerisation reactions is very low and that nearly bulk conditions prevail. As soon as this is confirmed experimentally, further research should concentrate on bulk polymerisation except for applications involving precipitation.

- Monomer evaporation behaviour at low gas saturation:
  All evidence points out that, without early build-up of a polymer hull, monomer will be lost for the largest part to the surrounding. When this prediction is confirmed, investigations involving gas without some degree of monomer saturation can be stopped completely, except for the special case of pre-polymerisation.

- Limitations of the monomer saturation in the drying gas:
  A higher monomer content in the gas both limits evaporation and drives (re)absorption from the gas, which helps to improve the yield. Yet, problems like condensation at cold spots or thermal polymerisation in the gas may arise and must be thoroughly analysed.

- Behaviour of reactive absorption:
  The mechanism of reactive absorption in conjunction with polymerisation is plausible, but so far just a prediction by the model with uncertainties in the underlying data. As this effect has a major impact on the overall process behaviour and yield, it needs to be analysed experimentally.

- Possibility of pre-polymerisation:
  The process window can be enhanced by partly polymerisation upstream

153

to the spray process. Challenges like clogging and spray generation of a polymer loaden solution need to be investigated in order to obtain the limitations of this process variant and its applicability.

Options for further model development are:

- Kinetic data:
  The kinetic model in this work was comparably simple and extrapolated from data at much lower monomer contents. More detailed data will improve the model accuracy.

- Thermodynamics and mass transfer:
  Absorption kinetics, which are currently calculated according to the same linear driving force as evaporation, can be refined. Additionally, instead of UNIFAC a more advanced model could be employed, which accounts for polymer peculiarities.

- Diffusion:
  The dependency of the diffusion coefficient on polymer content was based on an artificial approach which enables easy variation of diffusion parameters. Prediction of a real process needs more accurate data. Moreover, a more advanced mechanism than pseudo-binary Fickian diffusion would represent various diffusivities of the species in the solution better. Yet, this will require reliable data on diffusion coefficients of all components at all relevant mixture compositions.

- Consideration of boiling:
  So far, boiling of the mixture is not regarded in the model. When it comes to process design, this point has to be taken into account. Yet, this can also be checked during post-processing the data.

5. SMOOTHED PARTICLE HYDRODYNAMICS AND ITS APPLICATION TO SINGLE DROPLET SLURRY DRYING

Typical single droplet drying models presume spherical symmetry, like the droplet polymerisation model discussed in the previous chapters. The continuum is modelled in a quasi-homogeneous way, possibly with the distinction between a crust/shell and a core. In the following, a drying model based on the mesh-free SPH approach will be introduced, which calculates drying and morphology determining processes on a detailed scale. The basic physical effects incorporated within the model are depicted in Figure 5.1. Heat and mass transfer to the surrounding ($\dot{q}^{\Gamma}$, $\Omega_{H_2O}^{\Gamma}$) and heat conduction ($\dot{q}_{cond}$) determine drying. The motion within the slurry is affected by surface tension ($\vec{F}^{LL}$) and wetting ($\vec{F}^{LS}/\vec{F}^{SL}$) forces as well as interaction between suspended primary particles ($\vec{F}^{SS}$).

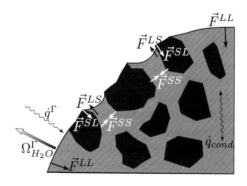

Figure 5.1: Physical effects incorporated in the SPH drying model

First of all, the SPH method itself will be introduced on a theoretical base. Thereafter, implementations of the various physical effects playing a role in droplet drying will be presented. As readers may have a process engineering/drying background with little knowledge on meshfree methods, the peculiarities of the method and challenges will be adressed, too. In the SPH literature the term "particle" is generally used for interpolation points, which can become confusing when drying of a slurry involving primary solid particles is to be modelled. The following explanations will be compliant with the common SPH wording. Suspended solids are called "primary particles". Otherwise the term "particles" refers to SPH points.

The model implementation considers the two-phase system of an incompressible Newtonian liquid and solid primary particles and their interaction by surface tension and wetting. Heat and mass transfer to/from the surrounding gas by linear driving forces is a new topic in SPH and will be derived in more detail. Within the droplet, heat conduction will be considered by an established approach. Furthermore, an extension of the mass transfer model to the second drying phase and different approaches for modelling crust formation will be introduced. As an alternative to linear driving forces and in order to underline the flexibility of the approach, diffusion driven drying of a porous structure will be derived via coupling of the SPH model with an underlying grid.

Droplet drying simulations have been carried out in two dimensions. Extension to three dimensions is possible in principle at substantially higher computational cost.

5.1 Mathematical Derivation

5.1.1 SPH Interpolation

Smoothed Particle Hydrodynamics can be derived from statistical interpolation theory (Lucy 1977). A quantity f at a position $\vec{x}$ within the domain Ω can be expressed using the following identity

$$\langle f(\vec{x}) \rangle = \int_\Omega f(\vec{x}') \, \delta\left(\left|\vec{x}' - \vec{x}\right|\right) d\vec{x}'. \tag{5.1}$$

Approximation of δ, the Dirac delta function, by a kernel function $W(r,h)$ with a length scale h and $\lim\limits_{h\to0} W(r,h) = \delta(r)$ leads to

$$\langle f(\vec{x})\rangle = \int_\Omega f(\vec{x}') \lim_{h\to0} W\left(\left|\vec{x}'-\vec{x}\right|,h\right) d\vec{x}' \tag{5.2}$$

or, applying a finite smoothing length h,

$$\langle f(\vec{x})\rangle = \int_\Omega f(\vec{x}') W\left(\left|\vec{x}'-\vec{x}\right|,h\right) d\vec{x}' + O\left(h^2\right). \tag{5.3}$$

The interpolation error is of second order as long as W is an even function (Colagrossi 2005, p. 23), which generally is the case in Smoothed Particle Hydrodynamics. In the SPH approximation, the integral is changed into a summation over a finite number of values associated to discrete volumes. Neglecting the error term the interpolated value at a position $\vec{x}_i$ becomes

$$f(\vec{x}_i) = \sum_j V_j f(\vec{x}_j) W\left(\left|\vec{x}_j-\vec{x}_i\right|,h\right), \tag{5.4}$$

with j running over the entirety of all interpolation points including i itself. In doing so, the infinitesimally small volume $d\vec{x}'$ is converted into the discrete volume V_j. This is the basic equation of the SPH method, employing a discretisation by a finite number of particles, each with a certain dedicated volume or mass. Considering the derivation, standard SPH is first order accurate. However, this statement about accuracy can rather be regarded as a rule of thumb or the average behaviour of the method, whereas numerical consistency in a strict manner is even not ensured for zero order functions (interpolation of a constant value does not necessarily yield this value again).

SPH particles may possess other physical quantities like temperature, pressure etc. as well. Despite the fact, that they typically represent a certain mass of the continuum, such particles may not be considered as granular objects (like billard balls), which is one of the great differences with respect to the discrete element method (DEM). SPH particles truly are interpolation points and the SPH discretisation truly approximates the continuum. This is depicted in Figure 5.2, in which the left frame shows the particle representation of the continuum via smooth, blurred interpolation points. In the SPH literature, particles are typically drawn as circles without blurring, which makes graphs easier to comprehend and will be used hereafter. In this way, the right frame shows the confined neighbourhood around a particle of interest (black) with a bell shaped kernel.

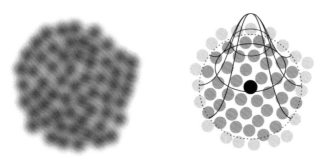

Figure 5.2: Particle representation of the continuum, left: smooth interpolation points, right: confined neighbourhood around a particle of interest (black).

For reasons of clarity and comprehensibility, the following, short expressions will be used hereafter: Values $f(\vec{x}_i)$ at a point i are expressed by f_i, the distance vector between point i and j by $\vec{r}_{ij} = \vec{x}_i - \vec{x}_j$ (this is reverse to the typical mathematical notation of distance vectors, but corresponds with common SPH literature) with an absolute value of $r_{ij} = |\vec{r}_{ij}|$ and kernel values with respect to r_{ij} by W_{ij}. If the volume is expressed by mass m and density ρ, the following, common SPH interpolation form is obtained

$$f_i = \sum_j \frac{m_j}{\rho_j} f_j W_{ij}. \tag{5.5}$$

The kernel function W has to fulfil several requirements:

- Its limit value for the smoothing length tending to zero has to be the Dirac delta function: $\lim_{h \to 0} W(r,h) = \delta(r)$.

- The integral over the kernel is normalised: $\int W(|\vec{x}' - \vec{x}|, h) d\vec{x}' = 1$.

- The kernel is symmetric: $W(-r, h) = W(r, h)$.

- $W(r, h)$ has to be continuously differentiable for at least one time.

- The kernel has a compact support, so that it becomes zero if the distance exceeds a certain cut-off radius.

The first three requirements concern the approximation itself, whereas the fourth condition affects the calculation of spatial derivatives, i. e. the discretisation of the nabla operator. In a pure theoretical sense, the fifth demand is not necessary, as SPH is derived by an interpolation over the complete domain. However, the number of interacting neighbouring nodes is to be limited in order to keep the computational effort reasonable. Hence, the summations are only performed with respect to a restricted neighbourhood, which is determined by the kernel's cut-off radius. The Gaussian with very good approximation properties is therefore seldomly used, as it only becomes zero for $r = \infty$ and has to be truncated beyond the cut-off radius. Very common are spline approximation of the Gaussian, namely the cubic (M_4) spline and the quintic (M_6) spline, which has been applied throughout this work:

$$W(r,h) = w_d \begin{cases} \left(3 - \frac{r}{h}\right)^5 - 6\left(2 - \frac{r}{h}\right)^5 + 15\left(1 - \frac{r}{h}\right)^5 & 0 \leq \frac{r}{h} < 1 \\ \left(3 - \frac{r}{h}\right)^5 - 6\left(2 - \frac{r}{h}\right)^5 & 1 \leq \frac{r}{h} < 2 \\ \left(3 - \frac{r}{h}\right)^5 & 2 \leq \frac{r}{h} \leq 3 \\ 0 & \frac{r}{h} > 3 \end{cases} \tag{5.6}$$

The normalisation constant w_d is $\frac{1}{120h}$, $\frac{7}{478\pi h^2}$ and $\frac{1}{120\pi h^3}$ in one-, two- and three-dimensional approximations, respectively (Price 2012).

It has to be clarified that mass and volume of SPH particles are only identical to a physical mass and a physical volume for threedimensional approximations. In a two-dimensional derivation $d\vec{x}$ denotes an infinitesimal small area element and just a scalar infinitesimal small distance if only one spatial dimension is considered. The corresponding "volume" of an SPH particle is hence a finite area or length, respectively, and the particle mass has the unit of measure $\left[\frac{kg}{m^{3-d}}\right]$ with d being the number of spatial dimensions. Though this notation appears unusual at first glance, it allows for general SPH expressions of model equations without distinguishing between approximations of different spatial order.

5.1.2 Integral Approximations

Typically the SPH approximation 5.5 is derived in the above-mentioned procedure and, by differentiation and some further manipulation, used in order to derive spatial derivative operators. Alternatively, an integral approximation is frequently employed. This concerns both operators for second order derivatives and the derivation of corrected SPH operators. The Taylor approximation in one spatial dimension x

$$f_j = f_i + \frac{\partial f_i}{\partial x}(\vec{x}_j - \vec{x}_i) + \frac{1}{2}\frac{\partial^2 f_i}{\partial x^2}(\vec{x}_j - \vec{x}_i)^2 + \ldots \tag{5.7}$$

is multiplied by the kernel W_{ij} or its derivative $\frac{\partial W_{ij}}{\partial x}$ and integrated over the domain Ω

$$\int_\Omega f_j W_{ij} dx_j = f_i \int_\Omega W_{ij} dx_j + \frac{\partial f_i}{\partial x}\int_\Omega (\vec{x}_j - \vec{x}_i) W_{ij} dx_j$$

$$+ \frac{1}{2}\frac{\partial^2 f_i}{\partial x^2}\int_\Omega (\vec{x}_j - \vec{x}_i)^2 W_{ij} dx_j + \ldots \tag{5.8}$$

$$\int_\Omega f_j \frac{\partial W_{ij}}{\partial x} dx_j = f_i \int_\Omega \frac{\partial W_{ij}}{\partial x} dx_j + \frac{\partial f_i}{\partial x}\int_\Omega (\vec{x}_j - \vec{x}_i)\frac{\partial W_{ij}}{\partial x} dx_j$$

$$+ \frac{1}{2}\frac{\partial^2 f_i}{\partial x^2}\int_\Omega (\vec{x}_j - \vec{x}_i)^2 \frac{\partial W_{ij}}{\partial x} dx_j + \ldots \tag{5.9}$$

The desired operator is then obtained by truncating the Taylor approximation at a certain term and by employing kernel properties like the anti-symmetry of the kernel derivative.

5.1.3 First Derivatives

Differentiation of equation 5.3 yields

$$\langle \nabla f(\vec{x})\rangle = \int_\Omega \nabla f(\vec{x}') W(|\vec{x}' - \vec{x}|, h) d\vec{x}', \tag{5.10}$$

which after integration by parts results in (Colagrossi 2005, p. 22)

$$\langle \nabla f(\vec{x})\rangle = \int_{\partial\Omega} f(\vec{x}') W(|\vec{x}' - \vec{x}|, h)\vec{n} dS' - \int_\Omega f(\vec{x}') \nabla_{\vec{x}'} W(|\vec{x}' - \vec{x}|, h) d\vec{x}'. \tag{5.11}$$

Inside the domain the contribution of the surface integral is negligible. Using the antisymmetry of the kernel derivative $\nabla_{\vec{x}'} W(|\vec{x}' - \vec{x}|, h) = -\nabla_{\vec{x}} W(|\vec{x}' - \vec{x}|, h)$, or in short $\nabla_j W_{ij} = -\nabla_i W_{ij}$, the basic SPH gradient operator follows:

$$\nabla_i f = \sum_j \frac{m_j}{\rho_j} f_j \nabla_i W_{ij}. \tag{5.12}$$

$\nabla_i W_{ij}$ is the kernel gradient with respect to particle i

$$\nabla_i W_{ij} = \frac{dW}{dr}\vec{e}_{ij} = \frac{dW}{dr}\frac{\vec{r}_{ij}}{r_{ij}}, \tag{5.13}$$

where $\vec{e}_{ij}$ is the unit vector between particles i and j and $\frac{dW}{dr}$ is the kernel derivative with respect to the inter-particle distance, which can be obtained analytically from the kernel function itself. Monaghan (e. g. 2005) introduced an additional function $F_{ij} = \frac{1}{r_{ij}}\frac{dW}{dr}$ to provide

$$\nabla_i W_{ij} = \vec{r}_{ij} F_{ij}, \tag{5.14}$$

which may be numerically advantageous, if $\frac{1}{r_{ij}}\frac{dW}{dr}$ is well defined for r_{ij} becoming zero in order to avoid divisions by zero. Divergence or gradient oparators of vectorial values $\vec{f}$ are simply obtained by applying the scalar or tensor product between $\vec{f}$ and the kernel gradient ∇W_{ij}, respectively.

Equation 5.12 indeed is the simplest SPH derivative operator, but the one with the least desirable properties. It is not even zeroth order consistent, because gradients of a constant function will not necessarily be computed to zero due to the distortion of the symmetry within a particle's neighbourhood using arbitrarily distributed particles. Moreover, it does not provide anti-symmetry concerning the transfer of conserved quantities within a particle pair, so that for instance the momentum transferred between two particles will be different depending on which particle is on the left and on the right hand side of the equation. Different operators can be obtained as follows. Introducing an arbitrary, differentiable function Φ and using the product rule $\nabla f = \frac{1}{\Phi}(\nabla(f\Phi) - f\nabla\Phi)$, equation 5.12 can be rewritten to (Monaghan 2005)

$$\langle \nabla_i f \rangle = \frac{1}{\Phi_i}\sum_j \frac{m_j}{\rho_j}\Phi_j(f_j - f_i)\nabla_i W_{ij}. \tag{5.15}$$

As can be simply seen from the term inside the brackets, operators modified in such a way are zeroth order consistent. The most common modifications are $\Phi = 1$ and $\Phi = \rho$:

$$\langle \nabla_i f \rangle = \sum_j \frac{m_j}{\rho_j}(f_j - f_i)\nabla_i W_{ij} \tag{5.16}$$

$$\langle \nabla_i f \rangle = \frac{1}{\rho_i}\sum_j m_j(f_j - f_i)\nabla_i W_{ij}. \tag{5.17}$$

The latter operator follows Monaghan's "second golden rule of SPH" (Monaghan 1992), rewriting formulae such that the density is inside the operators (typically the nabla operator). Still, both operators are not conservative. This desirable property can be obtained by $\nabla f = \Phi' \left(\nabla \left(\frac{f}{\Phi} \right) + \frac{f}{\Phi'^2} \nabla \Phi' \right)$, again with an arbitrary, differentiable function Φ' and applying the chain rule. $\Phi' = 1$ and $\Phi' = 1/\rho$ yield

$$\langle \nabla_i f \rangle = \sum_j \frac{m_j}{\rho_j} (f_j + f_i) \nabla_i W_{ij} \tag{5.18}$$

$$\langle \nabla_i f \rangle = \rho_i \sum_j m_j \left(\frac{f_j}{\rho_j^2} + \frac{f_i}{\rho_i^2} \right) \nabla_i W_{ij}, \tag{5.19}$$

both being anti-symmetric and conservative within a particle pair as will be shown later concerning the discretisation of the momentum balance. On the other hand, these operators are not zeroth order consistent for obvious reasons. Depending on the application, appropriate discretisation operators therefore have to be carefully chosen. Still, the approximation quality of SPH is comparably poor, anyway, and conservation is one of the remarkable and desirable properties of this method. As a rule of thumb, one should therefore rather decide in favour of conservative operators than of zeroth order consistency.

5.1.4 Laplace-Operator and Divergence of Diffusive Fluxes

In principle, a Laplacian can be expressed by differentiating equation 5.12

$$\Delta_i f = \sum_j \frac{m_j}{\rho_j} f_j \nabla_i^2 W_{ij} = \sum_j \frac{m_j}{\rho_j} f_j \frac{d^2 W_{ij}}{dr^2}. \tag{5.20}$$

Such formulations based on the second kernel derivative have turned out to be very sensitive to particle disorder. Moreover, depending on r_{ij}, the sign of the kernel's second derivative changes, which may cause unphysical effects like a heat flux from cold to warm regions (Monaghan 2005). Brookshaw (1985) obtained an alternative formulation by an integral approximation similar to equation 5.9, where the Taylor series is divided by $x_j - x_i$ before integration:

$$\int_\Omega \frac{f_j - f_i}{x_j - x_i} \frac{\partial W_{ij}}{\partial x} dx_j = \frac{\partial f_i}{\partial x} \int_\Omega \frac{\partial W_{ij}}{\partial x} dx_j + \frac{1}{2} \frac{\partial^2 f_i}{\partial x^2} \int_\Omega (x_j - x_i) \frac{\partial W_{ij}}{\partial x} dx_j + O\left(h^2\right) \tag{5.21}$$

$$\frac{\partial^2 f_i}{\partial x^2} = 2 \sum_j \frac{m_j}{\rho_j} \frac{f_i - f_j}{x_i - x_j} \frac{\partial W_{ij}}{\partial x} = 2 \sum_j \frac{m_j}{\rho_j} \frac{f_i - f_j}{r_{ij}} \frac{\partial W_{ij}}{\partial x}. \tag{5.22}$$

Due to the anti-symmetry of the kernel derivative, the first integral term on the right hand side is zero. The second integral reduces to unity, as it contains the SPH derivative in integral form of a linear function with gradient one. Brookshaw (1985, 1994) points out the interesting fact, that, using the cubic spline kernel and a uniform particle distance equal to the smoothing length h, this onedimensional Laplacian reduces to the well-known finite difference formulation $\frac{\partial^2 f_i}{\partial x^2} = \frac{f_{i+1} - f_i + f_{i-1}}{h^2}$. This equality, however, only holds in one spatial dimension and for this special particle / smoothing length configuration. The Laplace operator can be generalised to arbitrary spatial dimensions to (Brookshaw 1994; Jubelgas, Springel, and Dolag 2004)

$$\nabla_i^2 f = \sum_j 2 \frac{m_j}{\rho_j} \frac{f_i - f_j}{r_{ij}} \frac{dW_{ij}}{dr} = \sum_j 2 \frac{m_j}{\rho_j} (f_i - f_j) \frac{\vec{r}_{ij} \nabla_i W_{ij}}{r_{ij}^2} \qquad (5.23)$$

$$\nabla_i^2 f = \sum_j 2 \frac{m_j}{\rho_j} (f_i - f_j) F_{ij}. \qquad (5.24)$$

The expression $\frac{1}{r_{ij}} \frac{\partial W_{ij}}{\partial r} \left(= \frac{\vec{r}_{ij} \nabla_i W_{ij}}{r_{ij}^2} \right)$ is well behaved (Brookshaw 1994). However, in a numerical computation the denominator becomes singular for two particles approaching the same position, so that often a formulation like $\frac{1}{r_{ij} + \eta_{div}} \frac{\partial W_{ij}}{\partial r}$ or $\frac{\vec{r}_{ij} \nabla_i W_{ij}}{r_{ij}^2 + \eta_{div}^2}$ is applied (e. g. Morris, Fox, and Zhu 1997), with a small constant η_{div} preventing division by zero. Apart from the pairing instability (see section 5.1.6), particles should not coincide in an incompressible liquid. The present numerical model drops the additional constant in order to use a precomputed inverse of r_{ij} without division by zero problems. Alternatively, equation 5.24 may be applied, where F_{ij} can be easily evaluated at $r_{ij} = 0$ for most kernel functions. Concerning numerical efficiency, this avoids divisions and is therefore advisable (cmp. appendix C).

Despite the Taylor series has been truncated before the third derivative, the approximation is only first order accurate, as the SPH kernel integral / summation is involved (cmp. section 5.1.1). Many physical effects necessitate the divergence of a diffusive flux rather than the Laplace operator. Such a flux is proportional to the gradient of a quantity f and a coefficient like the heat conductiviy, diffusion coefficient or viscosity in diffusive heat, mass or viscous fluxes, respectively. In that case, Brookshaw obtained the form

$$\langle \nabla (K \nabla f) \rangle_i = \sum_j 2 \frac{m_j}{\rho_j} \frac{K_i + K_j}{2} \frac{f_i - f_j}{r_{ij}} \frac{dW_{ij}}{dr} = \sum_j 2 \frac{m_j}{\rho_j} \bar{K}_{ij} \frac{f_i - f_j}{r_{ij}} \frac{dW_{ij}}{dr}. \qquad (5.25)$$

$\frac{K_i+K_j}{2}$ is the average coefficient $\bar{K}_{ij}$ in the flux relation. This arithmetic averaging follows directly from the Taylor approximation. As an alternative, the harmonic mean $\bar{K}_{ij} = 2\frac{K_i K_j}{K_i+K_j}$ is frequently applied, particular for heat conduction phenomena, where it was justified by Cleary and Monaghan (1999) from considerations with finite differences. Generally, a harmonic average is more advantageous at strongly discontinuous coefficients K, as for K tending to zero at one of both particles the average coefficient and hence the flux become zero, whereas an arithmetic mean has the limit value of dividing the larger K in half.

The Laplace operator and equation 5.25 can be regarded as the SPH divergence of a finite differences flux. The flux $\vec{j} = -K\nabla f$ between particles i and j in a finite difference approximation is $\vec{j}_{ij} = -\bar{K}_{ij}\frac{f_i-f_j}{r_{ij}}\frac{\vec{r}_{ij}}{r_{ij}}$. Taking this expression as a central difference operator, the flux is located at the midpoint $\vec{x}_{ij} = 0.5(\vec{x}_i+\vec{x}_j) = \vec{x}_i - 0.5\vec{r}_{ij}$. The typical diffusive contribution in transport equations is calculated according to the negative divergence of such a flux. An SPH divergence over these fluxes between i and its neighbours has to be taken with respect to the midpoints $\vec{x}_{ij}$. However, this does not require a completely new SPH approximation, but can be derived with respect to the original neighbours j. The cut-off radius and the smoothing length are divided in half for this SPH divergence because of the midpoint positions. The volumes of the midpoint particles correspond with those of the original neighbouring particles j multiplied by $\left(\frac{1}{2}\right)^d$. The associated kernel derivatives are the original values $\nabla_i W_{ij}$ multiplied by 2^{d+1}. The divergence of the fluxes is hence $\sum_j 2\frac{m_j}{\rho_j}\vec{j}_{ij}\nabla_i W_{ij}$, which yields the Brookshaw approximation for the Laplace operator (cmp. equations 5.23 and 5.25). This may appear as an academic consideration, but in fact allows for an easy implementation of zero flux Neumann conditions at open/free surface boundaries. Such conditions are required in various applications like diffusion in a liquid or adiabatic boundaries. Demanding the flux $\vec{j}$ over a discontinuity to be zero implies for SPH particles i and j of different phases $\vec{j}_{ij} = 0$ or $f_i = f_j$. As a zero flux does not contribute to the SPH summation in the Brookshaw operator, this is equivalent to the particle j not existing at all. In other words, this kind of Neumann condition is automatically achieved, if the summation is only carried out over neighbours with a non-zero flux, i. e. particles within the same phase as i. A free surface therefore naturally employs a zero flux condition like an adiabatic boundary without consideration of the gas phase. This will be of further importance, when heat and mass transfer over an interface are considered by averaged linear driving force approaches (section 5.7.2).

The alternative approach of applying first order SPH derivatives twice, an SPH divergence of an SPH gradient, involves the same deficiencies like numerical oscillations as in grid based methods when enforcing the same operator on the same set of nodes twice (a second derivative by taking two times a centred finite difference will only refer to every other node). Cummins and Rudman (1999) tested both variants and observed numerical oscillations when solving a Poisson equation discretised by subsequent SPH differentiations, whereas a smooth solution was be obtained by the combined operator 5.25. A second issue is sensitivity to particle disorder (Cleary and Monaghan 1999).

5.1.5   General Second Derivatives

Arbitrary second derivatives can again be obtained by integral approximations (Español and Revenga 2003; Monaghan 2005) with the final SPH operator

$$\left\langle \frac{\partial^2 f}{\partial x^\alpha \partial x^\beta} \right\rangle_i = \sum_j \frac{m_j}{\rho_j} \left[\kappa_{\partial 2} \frac{\Delta x^\alpha \Delta x^\beta}{r_{ij}^2} - \delta_{\alpha\beta} \right] \frac{f_i - f_j}{r_{ij}^2} \vec{r}_{ij} \nabla_i W_{ij}. \qquad (5.26)$$

The factor $\kappa_{\partial 2}$ is determined by the dimensionality of the problem and set to 4 (two-dimensional approximation) or 5 (three dimensions).

5.1.6   Choice of Kernel, Smoothing Length and Cut-off Radius

The SPH derivations, e. g. equation 5.3, provide an approximation error coupled with the smoothing length, not with the particle size. Introducing smaller particles (decreasing l_0) at constant smoothing length does not change the error term in principle. Yet, the larger number of particles will refine the resolution and introduce a smoother representation of the field variables. Vector directions, e.g. surface normals are more accurate, if a larger number of particles has been taken into account. Two different kernel functions, each with the same smoothing length, but a different neighbourhood radius, will therefore exhibit different accuracies in practice. These smoothing effects may be less prominent when the neighbourhood is further enlarged. Yet, in practice one aims to limit the amount of neighbouring particles for computational reasons (and in order to avoid the pairing instability, see below).

The SPH approximation strongly relies on the kernel function and its properties. From a mathematical point of view, a Gaussian exhibits excellent approximation features (Price 2012). However, it does not provide compact support and has to be truncated at a certain cut-off radius, which offsets its theoretical advantages. Typical kernel functions are spline approximations of a Gaussian within a certain cut-off radius or other kernels with compact support like the Wendland ones. One issue concerning SPH is its lack of mathematical understanding and formalism (the first of the SPH "Grand Challenges", defined by the SPHERIC communtiy, Vacondio et al. 2020), especially when particles become disordered. Many comments on kernel functions and the choice of the smoothing length h are therefore not based on exact mathematical derivation but rather on experience with respect to various applications. One example is the definiteion of h and its ratio to the cut-off radius. The smoothing length provides a scale, on which a quantity is smoothed out when using the interpolation formula 5.3, and determines the accuracy. Moreover, the particle spacing/size l_0 cannot be chosen independently from this value. Nevertheless, as Dehnen and Aly (2012) pointed out, there is no appropriate definition of h in terms of the smoothing kernel. It may be either a kernel's standard deviation σ, the inflection point (maximum of $|\nabla W|$) or the ratio $\frac{W}{|\nabla W|}$ at the inflection point. Besides the Gaussian, where all these values coincide, these various definitions of h are not unique for arbitrary kernel functions. A sound determination of the smoothing length would be subject to advanced studies by itself and shall not be discussed further. Still, this kernel property appears in practically every SPH publication, whereas its implications and even the definition remain partly unclear.

Especially if a kernel with mostly Gaussian behaviour is desired, higher order function have superior properties. Still, h cannot be chosen independently from the particle spacing l_0. Monaghan (2005) states that the smoothing length is typically chosen close to the particle spacing and points out that approximations were more accurate in case of $h = l_0$ or $h = 2l_0$ instead of $h = 1.5l_0$ (Monaghan 1992). Yet, the variety of kernels being employed in SPH was small at that time and these remarks will mostly apply to spline kernels, in particular the cubic spline. Another issue is the so called pairing instability, which especially occurs at high h/l_0 ratios. According to Price (2012) the cubic and other B-spline kernels are prone to complete merging in case of $h \gtrsim 1.5l_0$ and exhibit an intermediate behaviour for $1.5 \gtrsim h/l_0 \gtrsim 1.225$. The reason for this is that spline kernels were designed with the primary intention to provide a good density estimate, whereas the calculation of a derivative was of minor interest. As the derivative of bell shaped kernels approaches zero for $r_{ij} \to 0$, repulsive terms will also become zero in case of coinciding particles.

In this work a cut-off radius of 3.1 times the particle spacing was chosen. This appeared as a lower limit for simulations of surface tension and wetting effects and for the calculation of surface normals in the CSF approach (at least with the implementations being used in the present model so far). A cut-off radius smaller than $3l_0$ reduces the number of neighbouring particles significantly so that orientation and length of surface normals may be erroneous. For $r_{cut} \approx 3l_0$ the cubic spline $2h = r_{cut}$ yields the unfavourable smoothing length of $h \approx 1.5l_0$. Hence, the quintic spline ($3h = r_{cut}$, hence $h \approx l_0$ in this case) was chosen throughout the simulations in this study.

The situation is somewhat different, if corrected SPH operators are applied. As they enforce a distinct order of accuracy and modify the original kernel and its derivative, the theoretical implications for ordinary SPH kernels do not apply. For instance, Bilotta et al. (2011) pointed out, that in theory three or four neighbouring particles are sufficient in order to provide a moving least squares (MLS) approximation, with an adequate number of about ten neighbours in practice.

Variable Smoothing Length vs. Constant Resolution

In grid-based methods the mesh size is often varied throughout the domain in order to adapt to local geometries and the intensity of physical effects. In SPH this can be done by a variable smoothing length. In doing so, regions with little motion and interactions can be discretised in a coarse way and other parts of the domain can be refined, when particle motion has to be calculated very accurately there. In the original field of SPH, astrophysics, this is a very reasonable approach. Engineering applications of SPH scarcely employ this procedure as it introduces additional dependencies - calculation of h and definition of appropriate averaged kernels etc. As drying is to be calculated in this work, a reduction of particle size due to evaporation and an adaption of the spatial resolution at the droplet's outer rim appears reasonable on first sight. Yet, arbitrary jumps in the resolution are not possible (like in grid-based methods, which only allow local variations of the mesh size to a certain degree). A particle undergoing evaporation may be much smaller than neighbouring ones, especially those of the solid phase. Just considering momentum transfer between two particles of individual mass differing by an order of magnitude, the light particle will undergo an acceleration ten times higher than the larger one. Stability criteria in time-integration would hence become too severe that a simulation could be undertaken within a reasonable time-scale. Accordingly, the topic of a variable resolution was dropped after first, instable tests and not followed further.

5.1.7   Correction of the SPH Approximation

The derivations of SPH operators are based on the assumption that the neighbourhood of an SPH particle is full and symmetric so that there are no inhomogeneities in the kernel approximation using neighbouring particles. In practice, both pre-requisites can be violated. At a free surface or a boundary, the particle distribution is neither complete nor symmetric. Consequently, the kernel summation is no longer normalised due to this particle deficiency. Moreover, the derivation of first order derivative operators as stated above is not valid anymore, because the surface integral in equation 5.11 does not vanish. This strongly affects the implementation of boundary conditions. The same errors can, in a minor intensity, be observed for the fluid bulk, if the symmetry is deteriorated due to particle disorder.

There are numerous remedies for the problem of particle deficiency at free surfaces in the literature. In general, two different, main approaches can be distinguished. For one thing, the neighbourhood of particles near the free surface can be artificially filled by the use of additional, virtual or ghost particles. This procedure approximately restores the normalisation of the kernel summation, if the void regions are filled appropriately, which can become a cumbersome task for arbitrarily shaped boundaries/surfaces. This procedure is further discussed in section 5.2. On the other hand the SPH approximation can be corrected in several mathematical ways. Modified SPH kernels and derivatives can either be obtained by some constraining conditions, which are related to the desired order of accuracy (for instance Bonet and Lok 1999), or by considering and correcting the terms, which are normally cancelled out in the derivation of derivative operators (like in the CSPM method of Chen, Beraun, and Carney 1999). Alternatively, new operators can be introduced, which are applied in a similar way to their SPH counterpart, but have a different mathematical foundation (for example MLS operators, Bilotta et al. 2011). Such approaches have not been introduced into the present model, but are an interesting option for further development, especially concerning the equations of motion at the free surface.

5.2 Implementation of Boundary Conditions

In finite differences, points may be located directly on the interface so that a condition like a gradient or a Dirichlet value can directly be imposed on a boundary node. The finite volumes method is based on cells and influxes can be imposed directly on the cell boundary. SPH particles are neither of both. They bear a mass and volume and thus represent a certain part of the continuum, which is attributed to a specific phase and not to the boundary, but do not exhibit a clear, individual cell interface. The issue of implementing boundary conditions appropriately without losing some of SPH's desirable properties has not been solved in general up to now and remains as the second of the "Grand Challenges" of SPH (Vacondio et al. 2020). The implementation of boundary conditions often corresponds with the question of particle deficiency and corrections applied there. An exhaustive overview on the subject of boundary conditions in SPH cannot be provided here, so that rather the basic concepts, which are relevant within this work, will be explained in the following. For further informations the interested reader is referred to the standard SPH literature and the aformentioned paper by Vacondio et al. (2020).

5.2.1 Ghost Particles

Artificial (ghost) particles re-establish full occupation of a boundary particle's neighbourhood and therefore overcome the particle deficiency. In the most straightforward case, a surface or boundary is flat and the virtual particles can easily be obtained by a reflection of inner particles at the boundary plane. Properties of these ghost particles are set according to the ones of original particles while taking the respective boundary conditions into account.

Depending on the surface/boundary geometry, the generation of ghost particles may become difficult as simple mirroring at curved surfaces will lead to dense or diluted particle distributions. In such cases an advanced boundary treatment might be necessary. However, for flat boundaries the ghost particle technique is an easy and robust method, which restores neighbourhood completeness and consistency of the SPH method at the same time.

Periodic Boundaries

The easiest implementation of periodic boundary conditions assumes an infinte domain by just setting a lower $x^{\alpha,L}$ and an upper value $x^{\alpha,H}$ for the respective cartesian coordinate α and copying the particles in the vicinity of one of these bounding values to the void region outside the corresponding other one

$$x_i^{\alpha,mirr} = x_i^\alpha + n_i^\alpha \left(x^{\alpha,H} - x^{\alpha,L} \right). \tag{5.27}$$

n_i^α is the normal towards the periodic boundary and points towards the fluid bulk ($n_i^\alpha = 1$ for i being in the vicinity of $x^{\alpha,L}$ and -1 otherwise for i being near $x^{\alpha,H}$). Particles trespassing one of these domain boundaries will be inserted at the other side according to the same rule. Depending on the intended application, single particle properties are modified so that either an infinite domain is achieved or influx and outflux conditions, for instance for a pipe.

Dirichlet Conditions

A Dirichlet implementation can be applied by mirroring particles at the boundary plane. Ghost particles' values are obtained by point reflection of corresponding values around the desired boundary value

$$f_i^{mirr} = f^{wall} + \left(f^{wall} - f_i \right) = 2f^{wall} - f_i. \tag{5.28}$$

The particle-averaged value at the boundary is then the desired value f^{wall}. One very frequent application of this rule concerns no-slip velocity boundaries, where f equals the velocity component tangential to the boundary plane $\vec{v}^t$ and $f^{wall} = \vec{v}^{wall}$. The same applies to the velocity component in normal direction. In case of a spatially fixed wall, the no-slip conditions simplifies to $\vec{v}_i^{mirr} = -\vec{v}_i$. Commonly, zero velocities/fluxes across the boundary are realised by invertion of normal vector components.

Alternatively, the Dirichlet value can be assigned directly to all mirrored particles. As this enforces discontinuous profiles in normal direction, this method is generally inferior.

Neumann Conditions

Similarly to Dirichlet conditions, a zero flux Neumann boundary condition works with mirrored ghost particles and is implemented most easily by directily copying the respective values from the original particle to its ghost counterpart. Due to symmetry, their corresponding gradients become zero directly at the surface and the Neumann condition is satisfied. One example would be a free slip boundary, where the velocity component tangential to the wall has to remain unchanged, whereas a Dirichelt condition is applied to its normal part.

5.2.2 Insertion of Boundary Conditions into SPH Equations

In some cases a boundary condition can be directly inserted into the SPH discretisation. This is the case for the $\nabla(K\nabla f)$-operator 5.25, which calculates the negative divergence of a flux $\vec{j}_{ij} = -\bar{K}\frac{f_j-f_i}{r_{ij}}$. If the flux between particles i and j is determined by a boundary condition, it should be possible in principle to exchange the term $-\bar{K}\frac{f_j-f_i}{r_{ij}}$ in formula 5.25 for the boundary particle pair $i \leftrightarrow j$ by the corresponding boundary flux. This has not been tested for general flux values throughout this work, but works for the special case of zero fluxes as described in section 5.1.4. Alternatively - or in combination with a zero-flux condition - fluxes can be imposed as additional source terms.

Implementing Boundaries by the Use of a Source Term

Interfacial fluxes can be considered as an additional source term in the differential equations of particles in the vicinity of a discontinuity. Area based fluxes need to be reformulated according to the mass-/volume based discretisation of particles. Such a volume reformulation can be undertaken in analogy to the CSF approach (Brackbill, Kothe, and Zemach 1992), the common procedure in computational fluid dynamics to account for surface tension forces. In principle, the transformation of area to volume based values is as a rule undertaken by approximating a sharp surface delta function with a smooth kernel, corresponding with a conversion of a sharp interface area into a diffuse interfacial volume. If an interfacial flux is considered by a source term, the flux over the interface in the ordinary diffusive SPH operators (typically the Brookshaw *div grad* operator) has to be zero. This procedure will be discussed in section 5.7.2 for the implementation of linear driving forces based heat and mass transfer.

Besides the novel approach for heat and mass transfer in this work, this principle course of action can be found considering point sources in heat conduction problems (Monaghan, Huppert, and Worster 2005), sorption of a surfactant at an interface (Adami, Hu, and Adams 2010a) or heterogeneous chemical reactions at a surface (Ryan, Tartakovsky, and Amon 2010).

Implementation into Corrected Operators

Chen, Beraun, and Carney (1999) derived their corrected second order CSPM derivatives from a Taylor series, in which the first order terms do not cancel out and have to be taken into account for reasons of consistency. Considering an SPH particle placed on the discontinuity, a Neumann boundary condition can hence be implemented into such a corrected operator by replacing the first order term with the appropriate boundary condition, as Chen, Beraun, and Carney have shown for a onedimensional heat conduction problem.

It might be possible to extend this procedure to the first order term in the Brookshaw operator, which vanishes within the bulk, but is erroneously ignored for surface particles. However, to the author's knowledge such an approach has not been undertaken yet.

5.2.3   Repulsive Forces as Hard Sphere Boundaries

In fluid dynamic problems, penetration of particles into a wall has to be avoided. Nevertheless, such issues may occur for several reasons like an incomplete neighbourhood of the particles being involved and therefore erroneous SPH calculations of fluid pressure. A simple way in order to avoid such undesired effects lies in the introduction of repulsive forces, comparable to a hard sphere potential (Monaghan 1994, 2005). Two particles getting very close will experience a strong repulsion, preventing penetration. With respect to incompressible liquids and e.g. simulation of sloshing in a basin, these forces are not necessary in the ISPH approach used in this work. Moreover, such forces are somewhat "particlish" and not based on transport equations. Forces from the atomic scale are transfered to the continuum scale. Still, this approach is very easy to implement. As the charm of particle methods is often connected to their ability to give a good approximation for a problem on which other methods fail, such boundary conditions may be applied as long as they do not alter the representation of the continuum. However, one has to be sure about the implications of introducing additional forces on the overall flow behaviour.

In fact, a strong repulsive part is present in the current implementation of surface tension forces, which employs an atomistic approach as well. Interparticle forces are prone to penetration of solids and therefore demand some repulsive contribution. This will be discussed in section 5.5 in more detail.

5.3 Hydrodynamics of an Incompressible Liquid in SPH

5.3.1 Continuity Equation, Density Evaluation

The continuity equation in Lagrangian based frame of reference

$$\frac{D\rho}{Dt} = -\rho \nabla \vec{v} \tag{5.29}$$

is typically discretised using an SPH derivative, like operator 5.16

$$\frac{d\rho_i}{dt} = -\rho_i \sum_j \frac{m_j}{\rho_j} \left(\vec{v}_j - \vec{v}_i \right) \nabla_i W_{ij} \tag{5.30}$$

or, following Monaghan (1989, 1994), operator 5.17

$$\frac{d\rho_i}{dt} = -\sum_j m_j \left(\vec{v}_j - \vec{v}_i \right) \nabla_i W_{ij}. \tag{5.31}$$

The latter equation has the advantage, that the densities of particles i and j, whose values may be erroneous, do not have to be considered and only the (typically constant) mass of particles j has to be taken into account. Besides these standard discretisations some authors enforce reformulations of the continuity equation 5.29, such as using the natural logarithm of the density $\frac{1}{\rho} \frac{D\rho}{Dt} = \frac{D \ln \rho}{Dt} = -\nabla \vec{v}$ or the ratio of the reference density to the current density $\frac{D \ln \frac{\rho^0}{\rho}}{Dt} = \nabla \vec{v}$ (Grenier et al. 2009).

Whilst solving the continuity equation in differential form leads to an initial value problem, which is integrated over time, the current density can be directly evaluated by the simplest use of the interpolation operator 5.5 with $f = \rho$

$$\rho_i = \sum_j m_j W_{ij}. \tag{5.32}$$

Equation 5.32 can be considered as an integral form of mass conservation and hence a discretisation of the continuity equation as well. Provided that the particle masses are constant, the continuity equation is automatically fulfilled (Benz 1990). In fact, the derivative of equation 5.32 with respect to time directly yields Monaghan's discretisation of the continuity equation 5.31, as the temporal kernel derivative is $\left. \frac{dW_{ij}}{dt} \right|_i = - \left(\vec{v}_j - \vec{v}_i \right) \nabla_i W_{ij}$. Equation 5.32 can be considered as the traditional way of calculating the density in SPH in astrophysics. Moreover, in the context of incompressible SPH (see section 5.4) it is the standard form of density evaluation, albeit there exist ISPH implementations using a differential form of the continuity equation as well.

The particle deficiency at boundaries and free surfaces introduces a density error in equation 5.5/5.32. A correction of this equation by use of a zeroth order consistent kernel, like the Shepard one ($\tilde{W}_{ij} = \frac{W_{ij}}{\sum_k \frac{m_k}{\rho_k} W_{ik}}$), will flatten the interpolated density to a constant value. This restores the particles volume $\frac{m_i}{\rho_i}$, but a flat density cannot be used in an incompressible scheme for the pressure evaluation. Corrective schemes therefore employ the derivative/divergence formulations above in combination with corrected kernel derivatives, see Bonet and Lok (1999) for WCSPH and Keller (2015) for ISPH. Yet, despite the particle deficiency at the surface, these schemes are seldomly applied. The impact of particle deficiency in the divergence operator acts as if the neighbourhood of a surface particle was filled with ghost particles of identical velocity. At least in lateral direction this conforms with a free slip boundary, which is appropriate for the free surface.

In multiphase problems with strongly varying (physical) densities, the masses m_j of equally spaced particles assigned to various phases differ the same way. With the neighbourhood of an interphase particle consisting of particles of different kind, equation 5.32 yields a smeared density field near a discontinuity, resulting in greatly varying particles volumes $\frac{m_j}{\rho_j}$ in SPH operators. In order to retain consistent SPH operators, the density evaluation by summation can be modified in the sense that only the particle alignment in the neighbourhood is taken into account by $\rho_i = \frac{m_i}{V_i^*}$ and $\frac{1}{V_i^*} = \sum_j \frac{m_j}{\rho_j} \frac{\rho_j}{m_j} W_{ij} = \sum_j W_{ij}$ (Hu and Adams 2006)

$$\rho_i = m_i \sum_j W_{ij}. \tag{5.33}$$

Comparing both principle methods of density evaluation, the continuity equation and the summation over the kernel values, the second one exhibits the above-mentioned edge effects, whereas the first one bears the disadvantage, that the particle number density and the fluid density or the volume occupied by particles are no longer coupled and equation 5.32 is violated (Benz 1990). This has to be kept in mind, when the density is evaluated as an initial value problem.

5.3.2 Momentum Balance

The momentum balance 2.5 in a Lagrangian reference frame is

$$\frac{D\vec{v}}{Dt} = -\frac{\nabla p}{\rho} + \frac{\nabla \tau}{\rho} + \frac{\vec{f}}{\rho}. \tag{5.34}$$

Depending on the application, different kinds of external forces $\vec{f}$ have to be considered with gravity and surface tension being the most common effects. These forces are considered in a volume specific way in units of N/m^3, i. e. in case of gravity $\vec{f}$ equals $\vec{g}\rho$.

Pressure Forces

A trivial SPH discretisation of the pressure-related acceleration $\left.\frac{D\vec{v}}{Dt}\right|_p = -\frac{\nabla p}{\rho}$ using operator 5.12 is neither symmetric nor even zeroth order consistent. One of the main advantages of SPH relies in its conservation properties so that typically the symmetric operators 5.18 and 5.19 are employed:

$$\left\langle -\frac{\nabla p}{\rho} \right\rangle_i = -\frac{1}{\rho_i} \sum_j \frac{m_j}{\rho_j} (p_j + p_i) \nabla_i W_{ij}, \tag{5.35}$$

$$\left\langle -\frac{\nabla p}{\rho} \right\rangle_i = -\sum_j m_j \left(\frac{p_j}{\rho_j^2} + \frac{p_i}{\rho_i^2} \right) \nabla_i W_{ij}. \tag{5.36}$$

Summation of the transferred momentum for both particles i and j using equation 5.35 yields:

$$m_i \left.\frac{D\vec{v}_i}{Dt}\right|_{p,i} + m_j \left.\frac{D\vec{v}_j}{Dt}\right|_{p,j} = \frac{m_i \, m_j}{\rho_i \, \rho_j} (p_j + p_i) (\nabla_i W_{ij} + \underbrace{\nabla_j W_{ij}}_{-\nabla_i W_{ij}}) = 0. \tag{5.37}$$

Due to the antisymmetry of the kernel gradient the momentum transferred within a particle pair is conserved. Thus, the total linear and angular momenta are conserved as well. Albeit these symmetric operators do not provide zeroth order consistency (a constant pressure field may not yield a zero pressure force as $\frac{m}{\rho} \neq const.$), conservation of momentum is typically considered as being more important in SPH applications. Discretisations 5.35 and 5.36 are hence commonly used in non-corrected SPH schemes. Bonet and Lok (1999) investigated the consistency of different implementations of the continuity equation and the pressure term with respect to the variational principle. They found that both the density evaluation by SPH summation / interpolation 5.32 combined with the pressure gradient 5.36 as well as the combination of the continuity equation in differential form 5.29 with the pressure term 5.35 are variationally consistent. Mixed combinations or the non-conservative zeroth order consistent operators cannot be justified from their variational considerations. It remains, though, unclear, how severe the impact of non-consistent combinations on the simulation result is, which after all are frequently used in the SPH literature. The present model combines the typical ISPH density calculation according to the summation equation 5.32 with the conservative formulation 5.36 and is therefore variationally consistent according to Bonet and Lok.

Corrected operators like the ones of Randles and Libersky (1996) or Chen, Beraun, and Carney (1999) are not symmetric in pairwise particle interactions, which may affect the conservation properties. This could be a reason for their rare application in flow problems. Yet, Bonet and Lok (1999) were able to show from the variational principle, that the total linear and angular momenta are preserved in their corrective scheme despite the loss of pairwise symmetry.

Last to mention, pressure forces are linked to the tensile instability in SPH. As pointed out by Adami, Hu, and Adams (2013), the lack of zeroth order consistency leads to spurious pressure gradients within a constant pressure field and numerical results change with different background pressures. In case of a very low background pressure, particle clumping because of attraction between neighbouring particles with effectively negative pressures leads to the tensile instability.

Discretisation of the Dissipative Term

Stress forces in general are expressed by the divergence of the deviatoric stress tensor (or the Cauchy stress tensor, if the pressure term is not considered seperately). The SPH implementation can be performed in analogy to the pressure term by the operators 5.18 and (which is the more common discretisation) 5.19:

$$\left\langle \frac{\nabla \tau}{\rho} \right\rangle_i = \frac{1}{\rho_i} \sum_j \frac{m_j}{\rho_j} (\tau_j + \tau_i) \nabla_i W_{ij} = \frac{1}{\rho_i} \sum_j \frac{m_j}{\rho_j} \left(\tau_j^{\alpha\beta} + \tau_i^{\alpha\beta} \right) \frac{\partial W_{ij}}{\partial x^\alpha}. \quad (5.38)$$

$$\left\langle \frac{\nabla \tau}{\rho} \right\rangle_i = \sum_j m_j \left(\frac{\tau_j}{\rho_j^2} + \frac{\tau_i}{\rho_i^2} \right) \nabla_i W_{ij} = \sum_j m_j \left(\frac{\tau_j^{\alpha\beta}}{\rho_j^2} + \frac{\tau_i^{\alpha\beta}}{\rho_i^2} \right) \frac{\partial W_{ij}}{\partial x^\alpha}. \quad (5.39)$$

With respect to conservation properties and consistency the same considerations hold as for the pressure term discretisation. Generally, the stress tensor is calculated beforehand according to the rheology of the medium. In case of an incompressible, Newtonian liquid the stress term simply is $\frac{\nabla \tau}{\rho} = \frac{\nabla(\eta \nabla \vec{v})}{\rho}$. Using the operator 5.25, Morris, Fox, and Zhu (1997) derived the form

$$\left\langle \frac{\nabla \tau}{\rho} \right\rangle_i^{Newt} = \frac{1}{\rho_i} \sum_j \frac{m_j \left(\eta_i + \eta_j \right) \vec{r}_{ij} \nabla_i W_{ij}}{\rho_j \left(r_{ij}^2 + \eta_{div}^2 \right)} \left(\vec{v}_i - \vec{v}_j \right) \tag{5.40}$$

with η_{div} set to $0.1h$. As mentioned above (section 5.1.4), it is advisable to replace the term $\frac{\vec{r}_{ij} \nabla_i W_{ij}}{r_{ij}^2 + \eta_{div}^2}$ by $\frac{1}{r_{ij} + \eta_{div}} \frac{\partial W_{ij}}{\partial r}$ or F_{ij}. It can be easily shown, that the linear momentum transferred within a particle pair $i \leftrightarrow j$ is conserved. However, the corresponding momentum vector and the inter-particle distance vector are not collinear, as the former one is determined by the velocity difference $(\vec{v}_i - \vec{v}_j)$. Angular momentum within a particle pair and the overall domain is hence not conserved.

An alternative formula for the Newtonian viscous term is based on the artificial viscosity, which was originally implemented by Monaghan and Gingold (1983) in order to dissipate numerical instabilities in astrophysical simulations. An adaptiation to Newtonian viscosity has been provided by Monaghan (2005) and Colagrossi et al. (2011)

$$\left\langle \frac{\nabla \tau}{\rho} \right\rangle_i^{Newt} = \sum_j m_j \kappa_\eta \frac{2 \eta_i \eta_j}{\rho_i \rho_j \left(\eta_i + \eta_j \right)} \frac{\vec{v}_{ij} \vec{r}_{ij}}{r_{ij}^2 + \eta_{div}^2} \nabla_i W_{ij}, \tag{5.41}$$

with κ_η depending on the dimensionality (8 in 2D and 10 in 3D). It has the desirable properties of vanishing for rigid body rotation and conserving linear as well as angular momentum. Nevertheless, the formulation of Morris, Fox, and Zhu is much more popular in the published SPH models of viscous, incompressible free surface flow. Its lack of conserving angular momentum seems not to be troublesome in industrial fluid dynamic problems (Monaghan 2005). The present model also employs the "traditional" equation 5.40. Equation 5.41 would be an alternative for further model development.

The various sub-formulations of the aforementioned discretisations in the literature involve different averaging procedures for the viscosity and the density (e. g. Shao and Lo 2003; Szewc, Pozorski, and Minier 2012) and do not apply to the present model, in which a constant viscosity is applied. Even implementations using a second order kernel derivative (Takeda, Miyama, and Sekiya

1994; Chaniotis, Poulikakos, and Koumoutsakos 2002) or subsequently employing first derivative SPH operators twice (Watkins et al. 1996) have been published. These formulations bear the shortcomings, which have been discussed in section 5.1.4, and should not be applied.

5.3.3 Weakly Compressible SPH (WCSPH)

The first application of SPH to incompressible liquid flows was performed by Monaghan (1994), who allowed a slight compression of the liquid and coupled pressure and density by the Tait equation

$$p = \frac{c_s^2 \rho^0}{\gamma} \left(\left(\frac{\rho}{\rho^0} \right)^\gamma - 1 \right) = p^0 \left(\left(\frac{\rho}{\rho^0} \right)^\gamma - 1 \right). \tag{5.42}$$

The reference pressure p^0 in this relation depends on the adiabatic exponent γ (for water ≈ 7), the reference density ρ^0 and the speed of sound c_s (Adami, Hu, and Adams 2013). The adiabatic exponent needs not to be chosen by means of physics and can be smaller in order to avoid large pressure peaks due to density errors. The density itself is typically evaluated in WCSPH using the continuity equation in differential form 5.31. The pressure is different from zero, if the current density differs from its reference value. As a consequence of a non-divergence-free velocity field, the density changes and a pressure is built up which counteracts the violation of incompressibility. The speed of sound is adjusted such that density variations are kept in a low range. With (Monaghan 2005)

$$\frac{|\partial \rho|}{\rho} \sim \frac{v^2}{c_s^2} = Ma^2 \tag{5.43}$$

the density deviations in a compressible flow depend on the Mach number $Ma = \frac{v}{c}$. Typically, a maximum density variation of 1 % is allowed, corresponding to

$$c_s = 0.1 \cdot v_{max}, \tag{5.44}$$

with v_{max} typically being the maximum velocity throughout the computation. Due to the decoupling of fluid density and particle number density, solution of the initial value problem is prone to large density errors. As a remedy, density filtering after each n-th time step by Shepard interpolation (Dalrymple and Rogers 2006) or a first order MLS kernel (e. g. Colagrossi 2005) or an additional diffusive term in the continuity equation (Antuono, Colagrossi, and Marrone 2012) are usually employed in order to smear out density peaks. Additional procedures in order to enhance the numerical stability of WCSPH are an artificial viscosity (Monaghan and Gingold 1983) and the so called XSPH method in order to avoid particle penetration (Monaghan 1989).

During model development, WCSPH had been tested as well. Yet, the maturity of the ISPH implementation as depicted in the next section was not achieved easily and would have required additional effort concerning a proper initialisation of the particle setup and the abovementioned stabilising procedures. Starting with a reliable WCSPH implementation, it is surely possible to implement the slurry drying model likewise at reasonable invest.

5.4 Incompressible SPH (ISPH)

The incompressible SPH enforces the well-known procedures of projection methods, which have been established by Chorin (1968). In short, the pressure is used in order to project an intermediate velocity, which is not divergence-free, onto a space of a divergence-free velocity field. A general classification of such implementations has been provided by Guermond, Minev, and Shen (2006). In the context of SPH, Cummins and Rudman (1999) were the first to enforce a projection method for pressure calculation instead of weakly compressible SPH. Shao and Lo (2003) adapted their approach for free surface flows in the application of coastal engineering and introduced the term Incompressible SPH (ISPH). The principle algorithm is the same for both, as well as for most of the published ISPH models, and consists in a modified predictor-corrector scheme. Firstly, all known right hand side values (viscous and external forces) in the momentum balance will be evaluated in order to obtain an intermediate velocity field $\vec{v}^*$ at intermediate positions $\vec{x}^*$ (predictor)

$$\vec{v}^* = \vec{v}^n + \Delta t \left(\frac{\nabla \tau + \vec{f}}{\rho} \right) \tag{5.45}$$

$$\vec{x}^* = \vec{x}^n + \Delta t \vec{v}^*. \tag{5.46}$$

The superscript n denotes the values at the formerly finished time-step n. In the corrector step this intermediate velocity is modified by the pressure term and the final particle positions are obtained (in this case by the trapezoidal rule)

$$\vec{v}^{n+1} = \vec{v}^* + \Delta t \frac{\nabla p}{\rho} \tag{5.47}$$

$$\vec{x}^{n+1} = \vec{x}^n + 0.5\Delta t \left(\vec{v}^n + \vec{v}^{n+1} \right). \tag{5.48}$$

Under the condition that the velocity at time-step $n+1$ has to be divergence free ($\nabla \vec{v}^{n+1} = 0$), a pressure Poisson equation (PPE) can be derived

$$\nabla \left(\frac{\nabla p}{\rho} \right) = -\frac{\nabla \vec{v}^*}{\Delta t}. \tag{5.49}$$

Either equation 5.49 is directly used in conjunction with an SPH discretisation in order to obtain the pressure (like in Cummins and Rudman 1999) or the continuity equation 5.29 is enforced to rest the incompressibility constraint on the density difference (Shao and Lo 2003)

$$\frac{\rho^* - \rho^{ref}}{\rho^{ref}\Delta t} = -\nabla \vec{v}^*. \tag{5.50}$$

The reference density can be either the physical density ρ^0, the initial density of a particle or the density from the previous time-step ρ^n. The latter is more frequently used and can be employed without further ado. If the physical density is used as reference and the density is calculated by the summation equation 5.32, this sum must match the physical density well at the initialisation of a calculation. A high initial density deviation can enforce a high pressure peak, which bursts the particles apart right at the beginning. Moreover, surface particles will always exhibit a lowered density according to the summation equation, which does not comply with the physical density. Yet, referring to the density of the previous time step does not prevent accumulation of density errors. Frequently, the divergence free criterion and the density deviation are mixed - either by summation (Aly, Asai, and Sonda 2013; Hu and Adams 2009) or in subsequent steps (Hu and Adams 2007).

Various discretisations of the pressure term in the PPE can be found in the SPH literature, which primarily differ with respect to the averaging of the density and possibly a symmetrisation of the resulting system of equations. Typically, the Brookshaw formulation 5.25 of the second derivative is employed in a form like

$$\left\langle \nabla \left(\frac{\nabla p^{t+\Delta t}}{\rho} \right) \right\rangle_i = \sum_j \frac{m_j}{\rho_j} \frac{4}{\rho_i + \rho_j} \frac{p_i - p_j}{r_{ij}} \frac{dW_{ij}}{dr} \tag{5.51}$$

The discretised PPE is a system of linear equations $A\vec{p} = \vec{b}$, which contains the pressures p_i and right hand sides of the PPE b_i (either based on the divergence of the velocity field or the density error 5.50) for all particles i. The coefficient matrix A is set as follows

$$A_{ii} = \sum_j \frac{m_j}{\rho_j} \frac{4}{\rho_i + \rho_j} \frac{1}{r_{ij}} \frac{dW_{ij}}{dr} \tag{5.52}$$

$$A_{ij} = -\frac{m_j}{\rho_j} \frac{4}{\rho_i + \rho_j} \frac{1}{r_{ij}} \frac{dW_{ij}}{dr}. \tag{5.53}$$

The system of linear equations can be treated efficiently by iterative solvers, typically preconditioned Krylov subspace solvers (Meister and Vömel 2008). Shao and Lo (2003) symmetrised the matrix A by a slight modification of the Brookshaw formulation in order to use the conjugate gradient (CG) method, which is restricted to symmetric, positive definite coefficient matrices. Nevertheless, the asymmetrical system in the formulation of Cummins and Rudman (1999) can be evaluated by other Krylov subspace methods, such as GMRES or BiCGStab (Meister and Vömel 2008). Generally, these solvers are known for very efficiently damping oscillations on a small scale in the mathematical solution and therefore to converge very quickly. In case of a bigger computational domain, oscillations may also occur on larger scales, and Krylov subspace methods become less efficient (the number of iterations grows with the size of the domain). This issue can be solved by adaptive multigrid solvers, which treat a problem on different spatial levels with varying resolution and therefore damp oscillations on various scales very efficiently. The PPE of large scale SPH calculations with many particles is therefore more efficiently treated by a multigrid method either for direct solution or alternatively as a preconditioner for a Krylov subspace method.

Unlike in grid-based methods, the pressure correction is typically applied only once and not iterated until the density error falls below a certain threshold value. An exception is provided by Hu and Adams (2007). One reason for this is the large computational effort for the solution of the PPE in SPH discretisation. Moreover, SPH is often used in free surface flow applications, where surface particles always exhibit a density error as their neighbourhood in equation 5.32 is incomplete. Another SPH peculiarity is that negative pressures may cause stability problems and are either avoided by clipping negative values after pressure correction to zero (Keller 2015) or superimposition of a reference pressure (Hu and Adams 2007).

5.4.1 Boundary Conditions in ISPH

One of the pleasant features of weakly compressible SPH is the fact, that one is not necessarily concerned with boundary conditions. As long as the density is evaluated in a sensible way, the pressure field relaxes to reasonable values as well. Solution of the linear system of equations in ISPH requires distinct, application-dependent boundary conditions. This subject is not given as much regard as one might expect in the SPH literature. Actually, Cummins and Rudman (1999) recommended the development of a more general boundary treatment in their pioneering work and not all of these issues have been solved. Many ISPH applications mainly refer to free surface flow and sloshing and are mostly concerned with wall boundary treatment and the free surface condition. It lies not in the focus of this work to resolve the issue of general boundary consideration within incompressible SPH, so that the interested reader may be referred to the ISPH literature. One possible start is the work of Hosseini and Feng (2011), who seize on the investigations of Guermond, Minev, and Shen (2006) and derive consistent pressure boundary conditions for incompressible SPH. Considering droplet drying, indeed the treatment of free surfaces is of particular interest.

5.4.2 Boundaries by the Ghost Technique, Wall Boundaries

Periodic boundaries can be represented by simply copying particles (see section 5.2). When PPE matrix entries of ghost particles are added to the positions of their reference particles and ghost entries are deleted from the matrix, the domain consists of a closed loop in direction of the periodic boundary. Similarly mirror boundary conditions can be applied, which was used by Cummins and Rudman (1999) in order to calculate inner flow problems bounded by a no-slip condition.

A different implementation of solid walls was undertaken by Shao and Lo (2003), who adapted an easy approach from the moving-particle semi-implicit (MPS) method (Koshizuka, Nobe, and Oka 1998). Solid boundaries are implemented by several layers of solid particles. The layer closest to the moving liquid is included into the pressure Poisson equation and their pressure values are copied to the layers being farther away from the liquid. In doing so, the neighbourhood of liquid particles on the wall side is completely filled and the pressure gradient within the wall is set to zero. This technique can easily be used in sloshing problems, like the dam break in a basin.

5.4.3 Free Surface Boundaries in ISPH

Most ISPH implementations treat the free surface boundary condition in a rather simple manner. Particles in direct contact with the free surface are detected by an appropriate method and then exposed to the Dirichlet pressure condition $p^{surf} = 0$ (Shao and Lo 2003; Lee et al. 2008). Corresponding publications hardly mention the detailed way of implementation, however, it is commonly stated that these particles are given $p_{i \in surf} = 0$. This implies a modification of the PPE such that this condition is directly imposed on such particles, setting the pressure of those particles zero "the hard way".

Keller (2015) proposes a different kind of free surface treatment based on ghost particles. For each free surface particle i all neighbouring inner fluid particles j are mirrored into the void region via a point deflection at this particle's centre (surface particles are left out). By this, completeness of the neighbourhood is mostly restored. Ghost particles have to be taken into account within the system of linear equations. Their virtual matrix entry A_{ij^*} is identical to the one of the original particle A_{ij}, because equation 5.53 does not depend on the direction of the particle distance but solely on its absolute value. Likewise, the contribution of ghost particle j^* to the main diagonal entry of the surface particle i is the same as from its original particle j. The sum of equation 5.52 needs just to be counted twice for all neighbours, which were mirrored. Imposing the condition that virtual particles have zero pressure $p_{j^*} = 0$, the PPE matrix needs not to be extended. Hence, simply doubling inner neighbours' contributions to the entries A_{ii} provides the condition of zero pressure at the free surface

$$A_{ii,i \in surf} = \sum_j \frac{m_j}{\rho_j} \frac{4}{\rho_i + \rho_j} \frac{1}{r_{ij}} \frac{dW_{ij}}{dr} + \sum_{j \notin surf} \frac{m_j}{\rho_j} \frac{4}{\rho_i + \rho_j} \frac{1}{r_{ij}} \frac{dW_{ij}}{dr}. \tag{5.54}$$

Bøckmann, Shipilova, and Skeie (2012) derive a very similar boundary treatment by multiplying the Dirichlet condition $p_{i \in surf} = 0$ with A_{ii} and a weighting constant C. Addition to the i-th line of the PPE leads to the modified entry

$$A_{ii}^* = (1 + C) A_{ii}. \tag{5.55}$$

Bøckmann, Shipilova, and Skeie (2012) set $C = 1$, which matches Keller's approach, apart from inclusion of neighbouring particles belonging to the surface. An (effectively) similar condition is derived by Nair and Tomar (2014).

The detection of surface particles is typically performed by a property, which differs between surface and bulk, and some threshold value. Typical properties are the density (Keller 2015; Shao and Lo 2003) and the divergence of the particle positions (Bøckmann, Shipilova, and Skeie 2012; Keller 2015; Lee et al. 2008). Additionally, the absolute value of the surface normal (see section 5.7.2) was tested throughout this work.

This way of free surface treatment involves some issues. Detection criteria may fail, for instance if surface layers are compressed due to surface tension forces. Moreover, due to their switching-type behaviour, a particle may be considered differently within following time steps - once as a zero pressure surface particle, then as a bulk particle, some steps later again as being part of the free surface. Thirdly, just imposing $p = 0$ on surface particles does not allow for any pressure correction between surface particles and does not take into account, that these particles still belong to the liquid and do not represent the infinitesimally thin interface. This mainly applies to the direct imposition on surface particles and not to the more relaxed modification of the PPE matrix of Bøckmann, Shipilova, and Skeie (2012) and Keller (2015).

5.4.4 Modifications to ISPH in This Work

Free Surface Condition

The zero pressure boundary implementations of Keller, Bøckmann, Shipilova, and Skeie and Nair and Tomar are considerable improvements in comparison to the naive implementation of strictly enforcing $p = 0$ for surface particles. They differ in detail, but can be considered as a variant of a more general concept, the use of penalty functions in order to restrict a degree of freedom (Bathe 1996, p. 143ff; Askes and Ilanko 2006). The linear system of equations is modified in such a way, that $A\vec{p} = \vec{b}$ becomes $\left(A + A^P\right)\vec{p} = \vec{b} + \vec{b}^P$, in which the superscript P denotes the contribution of the penalty function. In a continuum formulation this can be expressed as (Escobar-Vargas, Diamessis, and Loan 2011)

$$\nabla\left(\frac{\nabla p}{\rho}\right) + \tau^P\left[condition\right] = -\frac{\nabla \vec{v}^*}{\Delta t}. \tag{5.56}$$

τ^P is the penalty coefficient and the (boundary) condition is placed inside the brackets. With growing τ^P the system is dominated by the constraining condition and the conventional part of the equation becomes negligible. Typically, $\tau^P \gg A_{ii}$ is set for equation i in the system of linear equations (Bathe 1996, p.

144). The weak imposition of the boundary condition with finite τ^P improves numerical stability (Escobar-Vargas, Diamessis, and Loan 2011). Imposing $p_i = 0$ in ISPH simply results in addition of τ^P to the diagonal entry A_{ii}, with $p_i \to 0$ for $\tau^P \to \infty$ ($b_i^P = 0$ in this case). The above-mentioned approaches utilise this method by adding a value of $\tau^P \approx A_{ii}$, a rather small penalty coefficient. Due to the relaxed penalty procedure, the pressure distribution at the surface is improved in comparison to strictly setting $p = 0$.

This observation allows for a more general and smooth imposition of the free surface boundary condition without the need of a switch function. The underlying idea is that for all particles a penalty function is introduced with τ^P depending on a quantity which becomes large in the vicinity of a free surface and vanishes for bulk particles (in a mathematical sense $\tau_i^P \sim f^{surf}\left(r_{i\leftrightarrow surf}\right)$ with $r_{i\leftrightarrow surf}$ being the distance of particle i to the surface and $f^{surf}\left(r_{i\leftrightarrow surf}\right)$ a positive function monotonically decreasing to zero). The benefit of such an approach is that all particles moving towards or away from a surface will experience a gradual change of τ^P. Thus, the zero pressure condition is smoothly imposed without undesirable side-effects of a switch function. Suitable quantities for such a condition are identical to the properties, which are used for surface detection in established schemes. The divergence of the particle position $\vec{x}$ is equal to the number of spatial dimensions d in the fluid bulk. For a particle directly located at a flat surface, it tends to $d/2$, as half of the neighbours' contribution in the discretisation formula is missing due to the particle deficit. The measure $1 - \frac{\nabla \vec{x}}{d}$ provides therefore the desired properties of the above-mentioned function $f^{surf}\left(r_{i\leftrightarrow surf}\right)$ without the need of evaluating $r_{i\leftrightarrow surf}$. The penalty coefficient used in this work is

$$\tau_i^P = A_{ii}\left(1 - \frac{\nabla \vec{x}}{d}\right)^2 \cdot C^\tau. \tag{5.57}$$

C^τ chosen to 4 provides a similar kind of penalty to particles at a flat surface as the approach of Keller (2015). Due to the power of two, its contribution vanishes fast towards the liquid bulk. Within this work, $C^\tau = 5$ was used. If outer particle layers of the liquid become compressed - which was an issue during model development of the surface tension and wetting approach (see section 5.5) -, $\nabla \vec{x}$ stays at comparably high values at the surface. Imposition of the zero pressure condition may necessitate C^τ values in the order of 100 in such cases. The penalty is only applied when $\nabla \vec{x} \leq d$. Otherwise, particles with an erroneously high $\nabla \vec{x}$ will experience a "negative penalty". This can be the case for subsurface layers, when the aforementioned layer compression occurs. The SPH term for evaluating the divergence of particle positions is (according to operator 5.16)

$$\nabla_i \vec{x} = \sum_j \frac{m_j}{\rho_j} (\vec{x}_j - \vec{x}_i) \cdot \nabla_i W_{ij} \qquad (5.58)$$

$$= -\sum_j \frac{m_j}{\rho_j} \vec{r}_{ij} \cdot \nabla_i W_{ij}. \qquad (5.59)$$

The second formulation is easy and efficient, when interparticle distances $\vec{r}_{ij}$ are stored in a Verlet list anyway (appendix C).

Incompressibility Constraint

Under surface tension forces outer fluid layers are compressed, which can cause significant density errors. This is not only true if the divergence of the intermediate velocity field (equation 5.49) is used as the right hand side of the PPE, but as well when the density deviation 5.50 is employed, with the density of the previous time step as reference. The physical density as reference cannot be applied to free surface particles, which exhibit a "natural" deficiency of density according formula 5.32. This issue can be circumvented when this density deviation is weighted with a function f^{bulk}, which is very low at a free surface and approaches one for bulk particles - just the opposite to the function f^{surf} of the penalty coefficient before. The PPE is then written as follows:

$$\nabla \left(\frac{\nabla p}{\rho} \right) = -\frac{\nabla \cdot \vec{v}^*}{\Delta t} - C^P f^{bulk} \frac{\rho^n - \rho^0}{\rho^0 \Delta t^2}. \qquad (5.60)$$

Other than in equation 5.50, the density ρ^n is not evaluated at the intermediate position $*$ but taken from the previous time step n. This modified PPE can thus be interpreted like this:

- The first term based on the divergence of the intermediate velocity $\vec{v}^*$ accounts for the incompressibility violation of the flow field at the predictor position.

- The second term does not refer to the intermediate values, but to the density error accumulated over prior time steps.

In this respect, the incompressibility constraint is not "counted twice", but the errors from previous time steps are removed additionally to the correction of the predicted velocity field. A similar concept is proposed by Aly, Asai, and Sonda (2013) and Hu and Adams (2009). The constant C^P determines the ratio or rate at which accumulated density errors are to be corrected in the bulk (where $f^{bulk} = 1$) per time step. Hence, the last term can be considered as the proportional contribution within a numerical density control loop. This concept can be extended to a numerical PI controller, if the density error is integrated over each time step. Whereas this proved to work for a flow with little particle motion, the integral contribution is prone to cause crashes when particles move within the particle collective. In such a case a particle might carry an accumulated integral density deficit to a flow region with perfect incompressibility. An implementation is however straightforward and can be investigated further by the interested reader. In the current model, the function f^{bulk} is set depending on the divergence of $\vec{x}$

$$f^{bulk} = \begin{cases} 1 + \nabla\vec{x} - d, & d - \nabla\vec{x} < 1 \\ 0 & \text{otherwise.} \end{cases} \tag{5.61}$$

The alterations to the standard ISPH procedure are easy to implement. The modified free surface condition only necessitates a modification of the PPE matrix A_{ii} entries based on $\nabla\vec{x}$ values. The proportional contribution contains values, which are present anyway, and is just added to the typical velocity divergence RHS. An integral term demands storage of an additional property (the integral density deviation), but is not more complicated than the proportional term otherwise.

Parameterisation in this Work

Drying simulations in this work were carried out with a proportional factor $C^P = 1$. This is necessary for modelling the receding of the liquid after crust formation correctly. Moreover, capillary pumping inside the crust can involve negative pressures inside the liquid. Other than often observed this did not cause instabilities in the droplet model, which may be linked to the addition of surface tension forces, but was not further investigated. Solution of the PPE was performed either by a preconditioned Krylov subspace solver from the PETSc library (Balay et al. 2019) or the BoomerAMG multigrid solver (Henson and

Yang 2002) of the Hypre library [1]. The AMG solver can work as a preconditioner to PETSc Krylov subspace solvers or stand alone. Throughout this work,
it was applied as the single solver, when particle numbers were above about
20000. Below this value, Krylov subspace methods were more efficient.

Possible Further Development

The abovementioned improvements are reasonable and provide a stable model
for slurry drying, especially with regard to the surface tension model depicted in
the next section. Yet, the parameters should be further analysed and compared to
standard ISPH in standard applications. The pressure calculation and evaluation
of pressure forces near the boundary could be improved by the use of corrected
SPH operators. To the author's knowledge, this has only been introduced into
ISPH so far by Keller (2015) using the corrections of Bonet and Kulasegaram
(2000).

Throughout this work a one step scheme was tested instead of the predictor
corrector approach. Assuming that the velocity field at step n is divergence free,
the right hand side of equation 5.49 equals $-\frac{\nabla \cdot \vec{v}^*}{\Delta t} = -\frac{\nabla \cdot (\vec{v}^n + \vec{a}^n \Delta t)}{\Delta t} = -\nabla \vec{a}^n$. The
divergence of the acceleration due to viscous, surface tension and gravitational
forces therefore should yield the pressure, if inserted into the PPE, without the
necessity of evaluating intermediate particle positions. Using this as the single
incompressibility constraint, this method diverges after a number of time steps,
as the effect of particle motion on the divergence of the velocity field cannot be
neglected. However adding proportional terms with respect to the accumulated
density error as above and the divergence of the velocity field $\vec{v}^n$ a stable solution
was obtained with proportional factors of 0.2 for each term. This approach was
able to finish a small scale drying simulation and appeared to consume 10 %
less runtime on a scale of 27500 particles. It requires further validation and
development and was therefore not used in productive simulations. Moreover,
this is just an ISPH equivalent of the explicit Euler method, which provides
only poor accuracy. Further development to a higher order method as leap-frog,
velocity verlet or multi-step methods should be possible though. This appears
attractive insofar that standard ISPH itself only employs a first order scheme
(Cummins and Rudman 1999).

---

[1] https://computing.llnl.gov/projects/hypre-scalable-linear-solvers-multigrid-methods

5.5 Surface Tension and Wetting

There are two common literature approaches for the implementation of surface tension and wetting phenomena into SPH. The first one is the Continuum Surface Force (CSF, also called Continuous Surface Force) method (Brackbill, Kothe, and Zemach 1992; Morris 2000), which basically is a continuous reformulation of the surface tension force on a singular interface. The second approach enforces a molecular point of view, in which the effect of surface tension and wetting is implemented by pairwise forces between SPH particles. Whereas the first method is sound and consistent and the standard procedure in CFD, it requires the consideration of both phases besides an interface and implies several demandings with respect to the particle setup near the surface (a free surface implementation was presented in the master thesis of Woog 2011, but involving strong smoothing, large neighbourhood radii and no wetting effects). The second one is very intuitive, easy to implement and can be applied in a free surface implementation as well. Nevertheless, it's theoretical foundation in the literature is mediocre and lacks a thorough discussion.

Drying of a slurry involves many three phase boundaries. Accounting for wetting phenomena using the CSF approach is possible (Hu and Adams 2006; Huber et al. 2016) when the gas phase is represented by SPH as well. In the current application, this would lead to a much higher computational overhead. Moreover, the CSF approach is based on a colour functions' values, which change between different phases and need to be differentiated by SPH operators. Due to drying, solid SPH particles will at some point be covered only by a thin layer of liquid ones, which makes the application of SPH operators on the colour function a very difficult task. SPH operators require smooth field variables, but this is not provided when the fluid phase consists of a thin layer. This holds all the more for a two-dimensional approximation of slurry drying as in the present model. When two solid primary particles approach each other during crust formation, the small channel between both particles becomes more and more narrow until it is closed, which will interfere with the calculation of a colour function. For these reasons, the CSF approach was not further pursued and interparticle forces were applied in all drying calculations within this work. It is, however, a weak spot in the current model and needs further development.

Possible Further Model Development to a CSF Approach

The matter of thin particle layers as a result of drying becomes less severe in three dimensions. The points of contact between primary particles are small so that a continuous colour field could be calculated within the remaining pore space. The above-mentioned overhead when implementing the gas phase could be alleviated by using an underlying grid instead of gas particles, which is solely used for the colour function's evaluation in the different phases (cmp. section 5.7.5). Still, narrow sections of the geometry will become challenging, when gaps are in the particle scale.

5.5.1 The Interparticle Force Approach

From a molecular point of view, surface tension results from the attractions between molecules (or atoms), which are often expressed by a Lennard-Jones potential. Indeed the surface tension coefficient can be derived from molecular dynamics simulation. Due to its simplicity, this concept has been adopted in particle methods on a much coarser scale in order to model surface tension and wetting forces. The basic features are depicted in Figure 5.3. Attractive forces between atoms or molecules in a liquid will cancel out in the fluid bulk (frame a), whereas the net force at the fluid surface will point towards the inner part of the droplet (frame b). This models a surface tension like behaviour. Yet, this force does not vanish on flat surfaces (frame c), whilst surface tension forces only appear on curved surfaces. The reason for this is that the interface is not sharp on the molecular scale, but provides a smooth transition with a diluted fluid density (frame d). This leads to pressure forces towards the interface (the bold arrow in frame d), which cancel out the normal component of the net force obtained by inter-atomic/inter-molecular attraction (Marchand et al. 2011).

An SPH implementation of this approach was provided by Nugent and Posch (2000) for a van der Waals fluid. The attractive (surface tension) forces evolve naturally from the cohesive pressure $-a\rho^2$ in a modified van der Waals equation of state

$$p = \frac{\rho \bar{k} T}{1 - \rho \bar{b}} - \bar{a}\rho^2, \tag{5.62}$$

in which the reduced quantities $\bar{k} = \frac{k_B}{m}$, $\bar{a} = \frac{a}{m^2}$ and $\bar{b} = \frac{b}{m}$ rely on Boltzmann's constant k_B and the parameters a and b from the common van der Waals EOS. Whilst Nugent and Posch obtained results in good agreement to the theory, their approach appears numerically demanding. The cut-off radius was considerably high (five times the particle spacing) and the cohesive pressure needed to be calculated seperately from the EOS with a neighbourhood of even doubled size. Moreover, this approach naturally involves the gas phase. Tartakovsky and Meakin (2005) took up the ideas of Nugent and Posch and mixed a van der Waals equation of state for pressure calculation in WCSPH with an artificial interparticle force

$$\vec{f}_{ij} = \begin{cases} s_{ij}cos\left(1.5\pi\frac{r_{ij}}{r_{cutoff}}\right)\frac{\vec{r}_{ij}}{r_{ij}}, & r_{ij} \leq 3 \\ \vec{0}, & r_{ij} > 3 \end{cases} \tag{5.63}$$

$$\left.\frac{D\vec{v}_i}{Dt}\right|_\sigma = \frac{\sum_j \vec{f}_{ij}}{\rho_i}$$

using a cut-off radius of three times the particle spacing (the original work uses $\frac{1.5\pi}{3}$ within the cosine, which does not provide the proposed attractive and repulsive behaviour). Despite the reduced neighbourhood they obtained stable and reliable results. In comparison to the work of Nugent and Posch (2000), the surface tension coefficient is not related to the cohesive pressure but to the pre-factor s_{ij}, which can be used as an independent parameter in formula 5.63. Dynamic simulations with oscillating drops showed a good agreement of the periodicity with the surface tension coefficient obtained by the Young-Laplace equation $p = \frac{\sigma}{R}$ from an equilibrated drop. The pressure could however not directly been taken from the equation of state, but needed to be obtained by summing up all pressure and interparticle forces within a "virial radius" (a centred, circular area within the droplet's bulk).

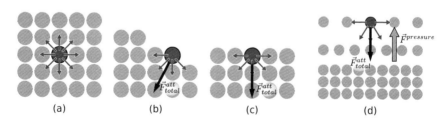

Figure 5.3: Surface tension by isotropic attractive forces between SPH particles: fluid bulk (a), curved surface (b), flat surface (c) and molecular scale (d), only interaction between nearest neighbours drawn.

When liquid-liquid (parameter s_{LL}) and liquid-solid (s_{LS}) interactions are parameterised with different strength, wetting effects can be simulated. Non-wetting behaviour corresponds with s_{LS} being much lower than s_{LL} ones with the limit value of zero attraction for a contact angle of 180 °. If, on the other hand, both values are equal, the solid is virtually treated as a liquid and the behaviour should be fully wetting. Kondo et al. (2007) derived both the magnitude of the interaction parameter inside a liquid as well as the liquid-solid parameter from an energy consideration. Supposing an energy potential $P(r)$ for the pairwise parti-

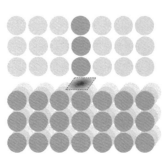

Figure 5.4: Detaching of a particle column from a flat substrate and surfaces being created (after Kondo et al. 2007).

cle interactions, the surface tension force is the derivative $f = \frac{dP}{dr}$. The surface tension coefficient is equal to the energy which is necessary to create one single surface from a fluid bulk. First of all, a flat surface discretised by particles is considered. A fluid column of one particle extent rests on this surface. This particle chain shares an interface area of l_0^2 with the surface. Separating the particle column from the surface (see Figure 5.4), the energy for generating two surfaces $2l_0^2\sigma$ is hence equal to the summation over all interactions between the particles in the column and the particles within the plane

$$2\sigma l_0^2 = \sum_{i \in I, j \in II} P(r_{ij}, s_{LL}). \tag{5.64}$$

An interaction parameter s_{LL} for liquid-liquid interactions in the potential P could hence in principle be calibrated by summation over the potential (not the force!) using an artificial geometry of such a fluid column. The potential provided by Kondo et al. (2007) yields, same as with Tartakovsky and Meakin, an attractive force for distances greater than l_0 and repulsion for smaller separations. The idea of a single particle column can be extended to wetting phenomena. If the surface does not consist of liquid, but solid particles, separation of both will generate a liquid and a solid surface at the cost of destroying a liquid-solid interface. The left hand side of equation 5.64 becomes $(\sigma_l + \sigma_s - \sigma_{LS})\, l_0^2$, whereas in the right part s_{LL} is exchanged by s_{LS} as interactions between liquid and solid particles are involved. A comparison of both formulae with Young's equation

$$\sigma_L - \sigma_{LS} - \sigma_L cos\theta = 0 \tag{5.65}$$

yields

$$s_{LS} = s_{LL} \frac{1 + \cos\theta}{2}. \tag{5.66}$$

Liquid-solid interactions can thus be calculated from the liquid-liquid ones and the contact angle θ with the limit values of s_{LS} being s_{LL} for $0°$ contact angle and $s_{LS} = 0$ in case of $180°$.

Simulations of Kondo et al. showed partly the intended wetting behaviour, but were unsatisfactory concerning the surface tension effect. Relaxation of an initially square drop does neither provide the intended oscillation period nor a relaxed state of a round droplet. Pictures of drops resting on a surface exhibit distorted surfaces. Kondo et al. attribute this to the differences of a regular particle chain resting on a flat surface to particles near a curved droplet surface with reduced interparticle distances in their simulations. An additional reason may be that the MPS method they used is similar to ISPH and hence does not employ an equation of state. The present ISPH model exhibits unwanted surficial particle alignments if the same force is used as by Tartakovsky and Meakin (cmp. section 6.4.3).

Discussion

Respresenting surface tension and wetting phenomena by simple, pairwise forces is very appealing. The implementation is easy to undertake and efficient. However, employing a molecular point of view on a continuum scale introduces some problems. The first issue is concerned with the parameterisation. The surface tension force or its prefactor s_{ij} needs to be calibrated with respect to the resolution and to the neighbourhood radius, as either a larger number of total particles or increase of neighbours per particle will result in a higher net force per unit area. The force is depending on the particle size and Adami, Hu, and Adams (2010b) pointed out that it does not converge to a limit value with the resolution being refined. This has not been addressed by the respective literature, so far. However, this issue can be partly circumvented, if the particle spacing l_0 is considered within the parameterisation. The net force per unit area grows with the number of particles per unit area, which is proportional to $\frac{1}{l_0^{d-1}}$ (d being the spatial dimensions in the simulation). The parameter s_{ij} can be therefore be calculated from a resolution independent interaction parameter s_{ij}^0 upon

$$s_{ij} = s_{ij}^0 l_0^{d-1}. \tag{5.67}$$

This is easily understood considering a surface in three dimensions. Doubling l_0 leads to a quarter of the total particle number contributing to the net force. The individual force per particle thus needs to be four times higher. In other words, the individual interaction parameter is the independent parameter times the individual surface area of the particle in a flat, Cartesian configuration (which is a line of length l_0 in 2D). This approach recalibrates surface tension forces, yet it fails at wetting phenomena, where the number of interacting particles in the vicinity of the contact line remains constant.

The second issue is of general nature. These forces also apply to the liquid bulk. As repulsive forces are applied below a distance of l_0, the particle alignment within the bulk is affected by pairwise forces, even if the net force just acts on surfaces. To the author's knowledge, the impact of innerparticle forces on flow properties of the bulk has not been studied, yet.

Thirdly, a particle located at a concave surface still experiences a net force towards the fluid bulk, whereas the real surface tension force points into the opposite direction due to the negative curvature. On the molecular scale, it is the pressure contribution from the decreasing density, which turns the net force away from the liquid bulk. This pressure contribution must be provided either from an equation of state (WCSPH) or from the ISPH pressure calculation. The calculated fluid pressure is therefore unphysically high, which is the reason why Tartakovsky and Meakin (2005) needed to sum up particle forces within a "virial radius" in order to relate the simulated pressure with the Young-Laplace equation. Also, the net force on a flat surface does not vanish. As long as the pressure field is only shifted to high values and pressure differences and accelerations are calculated the same as without the additional pressure, computational results remain unchanged. Yet, simulations may be dependent on the background pressure (Adami, Hu, and Adams 2013). Moreover, these unnatural forces can alter the surficial particle alignment and may not be compensated by opposite pressure contributions on the small scale near the free surface. As the approximation errors due to the particle deficiency at the surface worsen such effects, a fully corrected ISPH approach as introduced by Keller (2015) might alleviate the problem. Moreover, particles tend to penetrate into a solid wall, if the incompressibility constraint is based on $\nabla \vec{v} = 0$ and not the deviation from the reference density is taken into account or a stronger repulsive contribution is introduced. Finally, resting drops do not exhibit the contact angles as set via equation 5.66.

5.5.2   The Concept of Surface-Lateral Particle Forces

The idea of pairwise forces will work better, if attractive forces are only acting lateral to a surface, i. e. only particles having the same distance from the surface encounter attraction. Within a flat plane, the net force becomes zero. Curved surfaces will experience a net particle force depending on the curvature and point towards the correct direction. The basic principle is depicted in Figure 5.5 for a force solely acting on the outmost particle layer. Such lateral forces represent the net forces from the atomic scale. While the idea is appealing, an SPH implementation is challenging, as a clear measure of the distance from the free surface is needed and the method must be robust when particles move towards the surface or the fluid bulk. The first of both questions is a similar task as the identification of surface particles in ISPH (see section 5.4.3), but with more inner particles taken into account. "Natural" values, which exhibit a strong gradient near the surface, are the density or the particle number density, the length of the surface normal from the colour function $|\vec{n}| = \left|\frac{\nabla c}{|c|}\right|$ (see section 5.7.2) and the divergence of particle position $\nabla \vec{x}$ (equation 5.58 or 5.59). Numerical experiments with the density and the surface normal failed, as distortions of the density field alter these values and a continuous transition of particles between bulk and surface was not possible.

The basic idea of an surface-lateral force based on the divergence of $\vec{x}$ is as follows. As $\nabla \vec{x}$ is equal to the number of dimensions, $d - \nabla \vec{x}$ is a measure for the distance to the surface. If a particle is located directly on a flat surface, $\nabla \vec{x}$ should yield $d/2$. Attractive forces can therefore be concentrated on the outmost particle layer, if each particle gets an intensity factor k_i depending on its proximity to the surface. Multiplication of both factors yields the overall

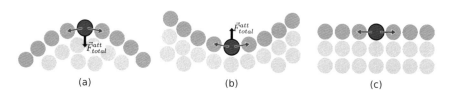

(a)   (b)   (c)

Figure 5.5: Basic concept of surface-lateral forces, attractive forces on a convex (a), concave (b) and flat (c) surface, only interaction between nearest neighbours drawn.

interaction intensity k_{ij}

$$k_i = 1 - \left.\frac{\nabla \vec{x}}{d}\right|_i \tag{5.68}$$

$$k_{ij} = k_i \cdot k_j. \tag{5.69}$$

Attractive interactions within the liquid bulk are thus inhibited and attraction of surficial particles with near bulk particles is strongly decreased. This approach yields the desired net force and is smooth enough to allow for a transition between bulk and surface. Yet, the force between the outmost layer and particles within the next inner layer is not zero, so that the second layer is wrongly attracted towards the liquid surface. The consequence is an accumulation of particles at the outmost layer. This effect can be diminished by strongly increasing the repulsive force for particles approaching each other. If this repulsion is restricted to particles interacting in the vicinity of the surface, bulk motion remains unaffected. While particle accumulation can be effectively prevented, this remains a solution of minor elegance and the net force of the second layer towards the surface still is a weak spot of this approach.

The procedure of calculating wetting effects can be derived by balancing the forces a liquid particle experiences in the vicinity of a three phase boundary, see Figure 5.6. A droplet resting on a plane is depicted on the left. The right frame shows the SPH representation in the vicinity of the contact line. The first force $\vec{f}_{LL}$ is simply the one within the liquid, parallel to its surface, directing away from the contact line (gray particles in SPH representation). The second one $\vec{f}_{LS}^{\parallel}$ is directed along the free surface of the solid, likewise away from the contact line (interactions of a surficial liquid particle with the solids with horizontal stripes). The third force $\vec{f}_{LS}^{\perp}$ is pointed in normal direction towards the solid (liquid particles interacting with both horizontally and vertically striped solids). By simple trigonometry, a liquid of contact angle θ remains at rest, if these

Figure 5.6: Forces acting at the contact line, left: continuum, right: particle representation.

forces cancel out by

$$\left|\vec{f}_{LS}^{\parallel}\right|\cos\theta + \left|\vec{f}_{LS}^{\perp}\right|\sin\theta = \left|\vec{f}_{LL}\right|. \tag{5.70}$$

The interaction intensities $k_{i,LL}$ and $k_{i,FL}^{\parallel}$ are both calculated according to equation 5.68. The contribution normal to the solid surface needs to take the overall surface of the solid into account, including the part covered by liquid. It is therefore calculated according to

$$k_{i,LS}^{\perp} = \frac{1 - \left.\frac{\nabla_{\vec{x}_S}}{d}\right|_i}{2}, \qquad \vec{x}_S \in S. \tag{5.71}$$

The factor $\frac{1}{2}$ needs to be applied as the influencing area of the solid stretches to both sides of the contact line and thus exhibits double the contribution compared to the other two forces. Interparticle forces are finally calculated according to

$$k_i = \begin{cases} 1 - \left.\frac{\nabla_{\vec{x}}}{d}\right|_i, & i \in L \\ \cos\theta \left(1 - \left.\frac{\nabla_{\vec{x}}}{d}\right|_i\right) + 0.5\sin\left(1 - \left.\frac{\nabla_{\vec{x}_S}}{d}\right|_i\right), & i \in S, \vec{x}_S \in S \end{cases} \tag{5.72}$$

$$f^{att}(r) = \cos\left(\frac{1}{2}\pi\frac{2r - (r_{cut} + l_0)}{r_{cut} - l_0}\right) \tag{5.73}$$

$$f^{rep}(r) = (r < l_0) \cdot f^{att}(r) \cdot k^{rep}\frac{l_0 - r}{l_0} \tag{5.74}$$

$$\vec{f}_{ij} = l_0 s_{ij}^0 \cdot k_i k_j \cdot \left(f^{att}(r_{ij}) + f^{rep}(r_{ij})\right)\frac{\vec{r}_{ij}}{r_{ij}}. \tag{5.75}$$

The distance dependence of the attractive interparticle force f^{att} is similar to formula 5.63 by Tartakovsky and Meakin, but extended to arbitrary cut-off radii so that at $r = l_0$ and at $r = r_{cut}$ the force becomes zero. A factor $k^{rep} = 500$ in the additional repulsive term was adjusted by numerical experiments and showed a satisfactory particle distribution. This repulsion becomes the stronger, the closer two particles get. One might consider a more hard-sphere-like behaviour with higher powers of $\frac{l_0 - r}{l_0}$, but this only provides stiff, strongly increasing forces at certain particle distances. A power of two led to a "pulsating" behaviour when particles were attracted, experienced strong, sudden repulsion and were attracted again. Repulsion needs to be stronger than attraction within a single particle pair as the latter acts on a larger number of long range neighbours and the particles of at least two outmost layers are "pressed" into a single layer. "Re-pairing" this behaviour by single pair repulsion requires strong forces in comparison to attraction. In numerical experiments with ISPH this holds for both the proposed surface-lateral force approach as well as the isotropic force suggested by Tartakovsky and Meakin. In WCSPH it may depend on the equation of state.

Implications on ISPH

A pairwise force approach is prone to spurious particle fluctuations which lead to a constant in- and outflux of particles to/from the surface, irregularly or in vortices. In such regions the density error is comparably high in the fluid bulk so that $\nabla \vec{x}$ may be calculated wrongly. Particle penetration can occur at the liquid-solid interface if the $\nabla \vec{x}$ field is distorted there. An additional incompressibility constraint based on the remaining density error (the proportional contribution discussed in section 5.4.4) alleviates this problem. The additional contribution can be relatively small. A proportional factor $C^P = 0.1$ for the density correction in the incompressibility constraint prevents unwanted effects. The overall density and pressure fields are satisfactory, yet with local peaks due to the "particlish" nature of pairwise forces which do not provide completely smooth values. "Traditional" isotropic forces necessitate a stronger additional density correction via higher C^P values when wetting phenomena need to be calculated. These observations are in accordance with Aly, Asai, and Sonda (2013), who implemented an alternative isotropic force into ISPH and provided an additional density contribution to the $\nabla \vec{v}$ constraint with a pre-factor of 0.25.

Interim Conclusion, Possible Further Model Development

The proposed new approach of in-plane tension forces is not fully developed yet. Net forces per particle are not completely lateral to the surface due to remaining attraction to inner layers. This disturbs the proper formation of a surface tension force. The current implemention lacks the ability to provide the same dynamic behaviour, when the particle size is altered. Moreover, scaling the force with the particle size normalises the net force on a free surface, but introduces an unwanted scaling of forces at contact lines. This will be shown for the validation test of wetting in section 6.4.4. Still, this approach yields a better droplet shape and representation of the wetting behaviour than the isotropic pairwise force method. The new approach is capable of approaching a correct (quasi-)static surface to a larger degree, which was a preliminary requirement for modelling drying. Due to the lack of scaling, the dynamic behaviour and capillary pumping will however be not represented equally on every particle scale.

One possible starting point for further investigations concerns the force. Both attractive and repulsive contributions are not entirely smooth. As Tartakovsky and Meakin (2005) pointed out, the force function can be arbitrarily chosen. Basic requirements are a long range attraction confined within the neighbourhood and repulsion for short distances. Further demand - at least according to the ISPH experience in this work - is that repulsive forces need to be significantly stronger than attractive ones. Aly, Asai, and Sonda (2013) smoothed the force from Tartakovsky and Meakin by multiplication with the kernel function W_{ij}. Besides smoothing, this strongly deminishes attractive forces due the bell shaped kernel, which can be an advantage concerning particle alignment.

As the anisotropic force exhibits a strong locality - the present approach mainly affects the outmost particle layer -, it violates the SPH necessity of smoothness to some degree. Smoothing the force field with a shepard kernel

$$\vec{f}_{ij}^{*} = \frac{\sum_j \vec{f}_{ij} W_{ij}}{\sum_j W_{ij}} \tag{5.76}$$

provides a more regular force distribution and may improve particle alignment. The reduced locality takes away dynamics and disturbs contact line motion to some degree. Yet, this alteration is easy to implement and can be adjusted by blending the localised force with its smoothed version.

The (at least partly) unwanted attraction between the outmost and the nextinner particle layer could be reduced, if the value $\nabla \vec{x}$ was calculated within a more restricted neighbourhood radius of e.g. two particles length. This would increase the locality of k_i and k_j values and concentrate the forces on the surface. Subsequent smoothing as mentioned before could then distribute this local force over neighbouring particles. The downside of such an approach is that the increase in locality goes along with a decrease of possible interaction partners.

Finally, the large repulsive part is yet necessary to prevent particles from forming "dense chains", but introduces additional artefacts. If particles are close enough, it can even flip the direction of the force on the outmost particles. An implementation of corrective terms, which restore the approximation capabilities of SPH at the droplet's surface, could alleviate the need for strong repulsion. Such an approach has been introduced by Keller (2015).

199

5.6 Representation of the Solid Phase

5.6.1 Primary Particles in the Slurry

Solid primary particles within a slurry are represented by SPH particles as well and need to be incorporated in a physically correct and at the same time efficient way. Implementation of the detailed material behaviour of the solid phase, e. g. as an elastic material, would introduce considerable computative overhead. Moreover, this is not necessary as material deformations of primary particles within a liquid are negligible. The solid phase is thus sufficiently represented by rigid bodies, which are allowed to move as collectives of single SPH particles in a translational or rotational way, but may not deform. The algorithm is simple and consists of two steps (Koshizuka, Nobe, and Oka 1998):

- SPH particles within a rigid body are treated as if they were belonging to the normal liquid. Their velocities are updated to temporary values according to pressure, viscous, surface tension and gravitational accelerations.

- Subsequently, linear and angular momenta of the whole body are summed up from the single SPH particles temporary velocities. These total momenta are redistributed amongst the particles to their final velocities such that the particle alignment is maintained.

The correction in detail is performed as follows:

$$\vec{x}_b^M = \frac{\sum_{j \in b} m_j \vec{x}_j}{\sum_{j \in b} m_j} \tag{5.77}$$

$$\vec{q}_{b,j} = \vec{x}_j - \vec{x}_b^M \tag{5.78}$$

$$I_b = \sum_{j \in b} m_j |\vec{q}_{b,j}| \tag{5.79}$$

$$\vec{T}_b = \frac{\sum_{j \in b} m_j \vec{v}_j}{\sum_{j \in b} m_j} \tag{5.80}$$

$$\vec{R}_b = \frac{\sum_{j \in b} m_j \vec{v}_j \times \vec{q}_{b,j}}{I} \tag{5.81}$$

$$\vec{v}_{b,j} = \vec{T}_b + \vec{q}_{b,j} \times \vec{R}_b. \tag{5.82}$$

$\vec{x}_b^M$ is the rigid body b's centre of mass. $\vec{q}_{b,j}$ is the displacement vector of a particle j belonging to body b with respect to its centre of mass and I_b the body's moment of inertia. $\vec{T}_b$ and $\vec{R}_b$ are the translatorial and rotational velocity vectors of body b, respectively. Within the ISPH calculation, this procedure is performed twice, once on the intermediate velocities $\vec{v}^*$ in the predictor and a second time after the final velocities have been obtained in the corrector step. In principle, the correction can be skipped, as the bodies shape is preserved after the corrector step. The additional rigid body evaluation should modify the pressure evaluation such that it is closer to the final rigid body motion. The benefit of the extra step is however not evaluated yet. As Koshizuka, Nobe, and Oka noted, the rigid body correction violates incompressibility to some (small) degree and is taken into account within the next pressure correction step.

Despite the abovementioned corrections, rigid body "blow-up" was observed when primary particles were undergoing rotation. The procedure only corrects rotational displacement for an infinitesimally small rotation. Within a finite time step, SPH particles are displaced with $\vec{v}\Delta t$ in a translational way, not on a circular orbit around the rigid bodies' centres of mass. Rotation therefore naturally leads to an enlarged distance of SPH particles from their respective centre. As a remedy, an additional correction was introduced after time integration:

- Calculation of new centre of mass $\vec{x}_b^{M,new}$ and displacement vectors $\vec{q}_{b,j}^{new}$

- If the ratio of $\left|\vec{q}_{b,j}^{new}\right|$ to $\left|\vec{q}_{b,j}\right|$ exceeds a certain threshold value, particle positions are corrected according to

$$\vec{x}_{b,j}^{corr} = \vec{x}_b^{M,new} + \frac{\left|\vec{q}_{b,j}\right|}{\left|\vec{q}_{b,j}^{new}\right|}\vec{q}_{b,j}^{new}. \tag{5.83}$$

This procedure is applied in each time step. It is not necessary to test all displacement vectors. As the "blow-up" happens isotropically, each pair of before/after displacement vectors yields, if the rigid body expanded unphysically. It just should be made sure, that the tested particle does not coincide with the centre of mass, which may occur in simple geometries. The present model allows a maximum expansion throughout the overall simulation time, with the condition

$$\frac{\left|\vec{q}_{b,j}^{new}\right|}{\left|\vec{q}_{b,j}\right|} - 1 > C^{RB,exp}\frac{\Delta t}{t_{max}}, \tag{5.84}$$

above which the aforementioned correction is applied. Present simulations used a $C^{RB,exp}$ value of 1 % expansion over the total computation.

Attraction between rigid bodies is calculated using the same pairwise force approach as for the surface tension calculation. Therefore, the factor s_{ij}^0 in equation 5.75 is multiplied with a prefactor f_{SS}, which calibrates the solid↔solid interaction in relation to the liquid attraction. Values k_i and k_j are calculated as if both particles belonged to the liquid as there is no contact angle between two solids. Due to the surface-lateral force approach, primary particles do only attract when they are not covered. This is physical insofar as the model is based on a macroscopic point of view, in which the attraction between two bodies is much smaller when separated by a liquid.

All rigid bodies bear an individual ID and can thus be distinguished. This ID is introduced as an additional particle property so that particle assignment to bodies is clear. In principle, this is sufficient for the execution of basic rigid body equations, if a very small number of bodies is to be evaluated. For an efficient treatment of a larger collective of bodies, a respective class was introduced so that each rigid body is represented as an object containing relevant data like the list of particles, their displacements to the centre of mass etc.

Possible Further Model Development

Rigid bodies are represented solely by SPH particles. Interactions between bodies are hence calculated on the scale of the involved SPH particles and depend on their local "degree of free surface" depending on $\nabla \vec{x}$. In the current implementation net forces between primary particles are resolution dependent. Rigid body interactions in the sense of discrete element (DEM) particles would improve their physical behaviour. The "blow up" correction could be avoided by direct calculation of particles' position from angular motion and the time step size. Still, the current correction may not be most elegant, but it works efficiently.

5.6.2 Calculation of Crust Formation

When the droplet's surface falls short of liquid, primary particles accumulate at the outer rim. At some point the droplet is covered by a solid crust. The basic principles of the implemented algorithms of crust formation are depicted in Figure 5.7. If no further treatment is appointed to the primary particles, the whole structure will shrink to a dense cluster of primary particles during further drying (frame a). The reason for this is that the forces between the solids are not strong enough to provide sufficient "stickiness" in order to form a stable crust.

The attraction between primary particles is relatively low in comparison to the drag exerted by the incompressible liquid. Additionally, attractive forces do not prevent rolling or sliding of primary particles. In order to simulate formation of a solid crust, primary particles are allowed to merge.

Caught on First Touch Method

The easiest implementation is to define a threshold distance, below which two primary particles merge (frame b in Figure 5.7). By this, primary particles agglomerate to a crust. The algorithm just involves a search over all particles within a rigid body whether their neighbours belong to a different body and undercut the limit distance. Sensible threshold values are in the order of the particle spacing l_0 or slightly below and could be used as an adjustable parameter.

Rigid body merging means that the SPH particles of both bodies get the same body ID and are collected within the one, single body object. The centre of mass, displacement vectors and the moment of inertia need to be recalculated for the newly created body. In order to track primary particle growth, it is reasonable to preserve the ID of the larger body.

Rigid Body Merging Depending on the Particle Environment

When bodies merge just depending on their distance, this is not necessarily related to drying. If two primary particles get close enough to each other within the droplet bulk, they will cluster according to the "caught on first touch" approach. Drying can be included, if the minimum distance condition is combined

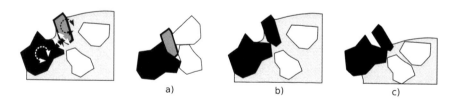

a) b) c)

Figure 5.7: Mechanisms of crust formation depending on primary particles' interactions, left: initial state, compaction without primary particle merging (a), "caught on first touch" (b), merging only when liquid fraction is low near the touching point (c).

with a minimum liquid content, below which merging may happen (see frame c in Figure 5.7). The liquid fraction of a particle i is obtained by SPH interpolation

$$\varphi_{L,i} = \sum_{j \in L} \frac{m_j}{\rho_j} W_{ij}. \tag{5.85}$$

The more liquid neighbours are in the vicinity of a solid SPH particle and the closer their distance is, the higher is the liquid fraction φ_L. Therewith a certain degree of drying in the vicinity of a solid particle is necessary to allow merging. It needs to be stressed that this is a local property of an SPH particle and does not account for the water content elsewhere at the rigid body. The threshold value $\varphi_{L,merge}^{lim}$ controls the amount of local drying being necessary for crust formation.

This concept can be extended to the solid content. It can occur that a spiky primary particle gets close to another solid in a comparably dry region. Both bodies will merge immediately, if the water fraction small enough, which acts as if the spike of one particle was glued on the other particle's surface. It appears reasonable that a certain solid content should be exceeded before merging takes place. The solid fraction φ_S is calculated the same way as above

$$\varphi_{S,i} = \sum_{j \in S} \frac{m_j}{\rho_j} W_{ij}. \tag{5.86}$$

Whereas the liquid fraction needs to undercut a value, the limit value $\varphi_{S,merge}^{lim}$ needs to be exceeded in order to allow body merging.

A high limit value of solid content and a low water threshold decrease the tendency to form stable solid clusters and delay or prevent crust formation. More relaxed limits involve earlier crust formation. In a physical sense these adjustable parameters be interpreted as the effect of a binder which increases the cohesion within a structure, which will be further explained in section 7.3.

Possible Further Development

A binder concentration as an additional particle property could relate primary particle merging directly to a physical value. Evaporation of liquid would lead to an increase in binder concentration. Transport of binder within the liquid could be modelled by diffusion. The stickiness of two primary particles could then be determined by their binder content. Additionally, the rigid body interaction might be investigated deeper and, as already mentioned above, be enhanced by DEM like mechanisms.

5.7 Modelling of Drying Phenomena in SPH

5.7.1 Heat Conduction

Heat conduction involves a term alike other diffusive effects, $\nabla (\lambda \nabla T)$, which can be discretised by the Brookshaw *div grad* operator 5.25. Brookshaw (1985) was also the first to introduce heat conduction into SPH. The corresponding energy balance in temperature notation is (Brookshaw 1985; Cleary and Monaghan 1999)

$$\frac{DT_i}{Dt} = \frac{2}{\rho_i c_p} \sum_j \frac{m_j}{\rho_j} \bar{\lambda} \frac{T_i - T_j}{r_{ij} + \eta_{div}} \frac{\partial W_{ij}}{\partial r}. \tag{5.87}$$

The averaged heat conductivity $\bar{\lambda}$ is an arithmetic mean, if derived from the integral approximation employing Taylor series (see section 5.1.4). In comparison, a harmonic mean provides the desirable feature that in case one particle has a heat conductivity of zero the transferred heat flux becomes zero as well (Cleary and Monaghan 1999). This allows for treating jumps of heat conductivities of three orders of magnitude accros three particle spacings (Monaghan 2005).

5.7.2 Implementation of Linear Driving Force based Heat and Mass Transfer into SPH

Linear driving force equations based on Nusselt and Sherwood numbers yield area specific heat and mass fluxes $\vec{q}^{\Gamma}$ and $\vec{\Omega}_j$ (equations 2.120 and 2.121). As has been stated before, a general course of implementation as in finite differences or finite volumes is not yet developed for the SPH method (section 5.2.2). Rather these area specific fluxes have to be implemented as additional source terms into the volume specific SPH discretisation. Hence, the individual surface area to volume ratio $\frac{A}{V}$ of a particle needs to be derived. This is an analogous problem to the consideration of surface tension forces, which are defined likewise on the interface between two phases and not in the fluid bulk. Typical treatment of surface tension in computational fluid dynamics is the continuous surface force (CSF) approach of Brackbill, Kothe, and Zemach (1992). The discontinuity of the interface is transfered into an interfacial volume of finite thickness d^{Γ}. Forces or fluxes, being interface area specific, are redistributed amongst this interfacial volume. The method is similar to the SPH derivation in section 5.1.1, in which the delta function around infinitesimally small points in the continuum is approximated by the finite kernel. The interface is infinitesimally thin in normal direction. Therefore, the CSF approach expresses the force acting on an interface area ΔA in a volume reformulation using an interface delta function δ^Γ

$$\int_{\Delta A} \vec{f}^A\left(\vec{x}^\Gamma\right) d^2x = \int_{\Delta V} \vec{f}^A \delta^\Gamma\left(\hat{n}\left(\vec{x}^\Gamma\right)\left(\vec{x} - \vec{x}^\Gamma\right)\right) d^3x. \tag{5.88}$$

The surface delta function is approximated by a weighting function $\wp$ ($\lim_{d\Gamma\to0} \wp = \delta^\Gamma$), which only changes in normal direction to the interface. Brackbill, Kothe, and Zemach define $\wp$ as the modulus of the gradient of a so called colour function c with a unit change over the interface. It is a smooth, gradually changing value within the interfacial volume of the CSF method and converges to a unit jump for a decreasing thickness of the interface. Likewise, $\wp$ converges to the delta function:

$$\lim_{d\Gamma\to0} \wp = \lim_{d\Gamma\to0} \left|\frac{\nabla c}{[c]}\right| = \delta^\Gamma \tag{5.89}$$

$[c]$ denotes the difference in the colour function over the interface. Division by this value guarantees that only a unit jump is considered and that the weighting function is normalised, independently from the choice of c. In principle, the colour function can be any value which changes between both phases, for instance the density (Brackbill, Kothe, and Zemach 1992). The surface tension force in a volume reformulation according to the CSF approach is then

$$\vec{f}^{\sigma,V} = \vec{f}^{\sigma,A}\left|\frac{\nabla c}{[c]}\right| \tag{5.90}$$

Application to heat and mass fluxes at the interface requires only exchanging the surface tension force with $\vec{q}^\Gamma$ and $\dot{\Omega}_j^N$. The interfacial volume of the CSF concept stretches on both sides of the interface. As the (normalised) colour function has a value of 0.5 directly at the discontinuity, $\wp$ contributes to the gas phase as well. If the gas was also considered by SPH particles, their share of heat and mass fluxes would need to be transferred to liquid neighbours. On the other hand, just half of the interfacial volume lies on the liquid side so that doubling their contribution makes this procedure obsolete

$$\dot{Q}^V = |\vec{q}^\Gamma|\, 2\left|\frac{\nabla c}{[c]}\right| \tag{5.91}$$

$$\dot{m}^V = |\vec{\Omega}|\, 2\left|\frac{\nabla c}{[c]}\right|. \tag{5.92}$$

The gradient of the colour function can be calculated in abscence of the gas phase by the SPH approximation of a property being one within the liquid phase and zero in the gas phase. The most simple SPH derivative operator 5.12 yields for a full neighbourhood including particles of the gas G (Woog 2011)

$$\nabla_i c = \sum_{j \notin G} \frac{m_j}{\rho_j} 1 \nabla_i W_{ij} + \sum_{j \in G} \frac{m_j}{\rho_j} 0 \nabla_i W_{ij}. \tag{5.93}$$

With the second summation being zero, the colour function's gradient is calculated exactly the same if only liquid (in the slurry drying model liquid and solid) particles are taken into account

$$\nabla_i c = \sum_j \frac{m_j}{\rho_j} \nabla_i W_{ij}. \tag{5.94}$$

The individual surface area to volume ratio of a particle in a free surface model is then

$$\left. \frac{A}{V} \right|_i = 2 \left| \sum_j \frac{m_j}{\rho_j} \nabla_i W_{ij} \right|. \tag{5.95}$$

Disturbances of the SPH particle alignment will lead to small fluctuations of this value within the liquid bulk. Therefore, this value is only taken into account if it exceeds a certain threshold value. Morris (2000) was the first to introduce the CSF approach into SPH for surface tension effects and applied a limit of $\left| \frac{\nabla c}{|c|} \right| > 0.01/h$. Smaller values are set to zero. The vector $\vec{n} = \frac{\nabla c}{|c|}$ provides the surface normal, which is further employed when surface tension forces are calculated, but not necessary here.

The additional source term in the temperature balance equation due to a linear driving force heat flux is finally

$$\left. \frac{DT_i}{Dt} \right|_{LDF} = -\frac{\alpha}{\rho_i c_{p,i}} (T_i - T^\infty) \left. \frac{A}{V} \right|_i. \tag{5.96}$$

The mass loss of a particle is

$$\left. \frac{dm_i}{dt} \right|_{LDF} = V_i \frac{MW_L}{\Re \bar{T}} \beta (p_{v,i} - p_v^\infty) \left. \frac{A}{V} \right|_i. \tag{5.97}$$

The interface is not sharp anymore, but stretches over particle layers of one cut-off radius. Drying will hence not only affect the outmost particle layer, but also to a lesser degree the second and, slightly, the third layer for $r_{cut} = 3.1l_0$. Similarly the heat flux is distributed. The cooling effect by evaporation is considered as an additional source term in the temperature equation

$$\left. \frac{DT_i}{Dt} \right|_{evap} = \frac{\Delta h_v}{V_i \rho_i c_{p,i}} \left. \frac{dm_i}{dt} \right|_{LDF}. \tag{5.98}$$

With Nusselt and Sherwood numbers provided, heat and mass transfer coefficients are calculated based on the characteristic length of $2R$ (equations 2.122 and 2.123). The droplet radius is detected at the beginning of each drying evaluation as the largest particle distance from the droplet's centre of mass.

5.7.3 Extension to the Second Drying Period

After accumulation of primary particles at the droplet's outer rim, only a small number of liquid particles is left at the droplet's surface and undergoes evaporation according to equation 5.97. Subsequently the droplet is heated up. Vapour transport through the porous crust is currently not considered in the model. An implementation of further evaporation in case of gas temperatures above the boiling point is however straightforward. When the liquid approaches the boiling temperature, additional heat being transferred to the droplet will directly be used for evaporation. This can be easily modelled by reversing equation 5.98

$$\left. \frac{dm_i}{dt} \right|_{boil} = \frac{V_i \rho_i c_{p,i}}{\Delta h_v} \frac{T^{boil} - T_i^*}{\Delta t}. \qquad T_i \overset{!}{>} T^{boil} \tag{5.99}$$

This is calculated in two steps. First the temperature equation is solved normally with respect to heat conduction and the linear driving force contributions (the implicit scheme is described in section 5.8.3). If the intermediate temperature T^* exceeds the boiling temperature, equation 5.99 is applied and the particle mass reduced likewise. Finally, T is set to T^{boil}.

As the crust falls dry, the droplet will exhibit free surfaces not only at its outer rim, but throughout the crust and at the surface of the receding liquid. In a naive implementation all these surfaces will exhibit a gradient of the colour function and experience heat transfer according to equation 5.96, which will exaggerate the heat input from the gas substantially. As a remedy, only particles at the outer surface are taken into account. The detection works as follows: The droplet is cut into wedges. Each particle can be unambiguously assigned to one wedge. The outmost particle of a wedge is considered as an interface particle. Subsequently, all neighbours of these outer rim particles are attributed as interface particles as well. The interface is thereby limited to at most four particle layers at the outer rim. The number of wedges must be chosen such that their outer arc length does not exceed the particle spacing l_0

$$n_{wedge} = \frac{2\pi R}{l_0} \tag{5.100}$$

in order to include inner layers completely in the second step.

Possible Further Model Development

For temperatures below the boiling point, a crust at the droplet's surface prevents further evaporation in the current implementation, as the mass loss is only dedicated to liquid particles. By implementation of vapour transport through the porous crust, further drying below the boiling point could be modelled and the transition to the boiling regime would become more gradual. The SPH formulation of diffusion is provided by equation 5.102. The diffusion coefficient needs to be chosen according to the effective permeability of the crust in this case.

5.7.4 Treatment of Evaporation Concerning Particle Mass and Deletion

Shrinking particles by reducing their mass m_i would effectively provide a collective of very small and comparably large particles, especially when a liquid particle is neighbouring a solid one. The matter of a variable resolution was shortly discussed in section 5.1.6 and is not a feasible approach to overcome this challenge. The SPH particle mass thus needs to remain unchanged. Evaporation is calculated on an additional particle property, an "evaporative" mass m^{evap}. This property is only attributed to drying and reduced until it approaches zero.

Particles of zero inner mass are deleted from the collective. The remaining gap in the particle configuration is closed by neighbouring particles within the following time steps due to incompressibility and surface tension effects. For reasons of conservation, the momentum of a deleted particle i is distributed amongst its neighbours j according to a Shepard interpolation

$$\vec{v}_j = \vec{v}_j^{old} + \vec{v}_i \frac{m_i W_{ij}}{m_j \sum_{k \neq i} W_{ik}}. \tag{5.101}$$

By this, linear momentum is preserved. Conservation of angular momentum is however slightly violated, but this is not visible in present simulation results.

Due to this stepwise drying behaviour of particle removal, a system at a relaxed state needs to equilibrate, again. In other words, energy conservation within the system is violated in such respect, that the potential energy is suddenly increased when the surface shape is altered by deletion of particles. As a result, the droplet is continuously relaxing from potential energy being locally introduced due to the discrete drying dynamics. This effect is less pronounced with decreasing particle size, as deletion of particles then occurs more often and the effects are more local. Still, some wobbling of suspended primary particles can be observed in simulations.

5.7.5 Modelling of Diffusion Driven Drying Involving the Gas Phase

Linear driving forces incorporate the interplay of diffusion and convection into an averaged transfer coefficient acting on a virtually linear profile over the boundary layer. Alternatively, mass and heat transport within the gas can be calculated directly by diffusion. In the following, a concept for diffusion driven drying will be laid out. Within this work, the approach was applied to drying of a flat, porous geometry.

Addition of the Gas Phase as SPH Particles

Evaporation due to diffusive transport between liquid and gas can be modelled directly, if the gas phase is modelled by an additional set of SPH particles. The diffusion terms can be implemented in a straightforward manner using equation 5.87 for the temperature derivation and, similarly, for the vapour pressure p_v

$$\frac{Dp_{v,i}}{Dt} = 2\sum_j \frac{m_j}{\rho_j} \bar{D} \frac{p_i - p_j}{r_{ij}} \frac{dW_{ij}}{dr}. \tag{5.102}$$

This is just an SPH discretisation of Fickian diffusion in the gas phase applying equation 3.13 to the vapour pressure ($\frac{\partial p_v}{\partial t} = \nabla(D\nabla p_v)$), which is possible if $c \approx const$. $\bar{D}$ denotes the average diffusion coefficient of vapour in the gas phase, calculated by a harmonic or arithmetic mean (see section 5.1.4). The diffusion induced velocity contribution is already considered in the diffusive operator (cmp. equations 3.7 and 3.13), so that the material derivative of this transport equation equals its Eulerian counterpart. The vapour pressure of liquid surface particles in equation 5.102 is set according to its local composition and the respective vapour-liquid equilibrium, for pure liquid the saturation pressure p_v^s. During time integration, liquid particles will encounter a virtual change in their vapour pressure. By use of the ideal gas law the corresponding vapour mass $m_{j,i}^v = c_{j,i}MW_jV_i = \frac{p_{j,i}MW_j}{\Re T_i}V_i$ corresponds with the particle volume (j denoting the component and i the particle). The local mass loss due to vapour diffusion can be calculated from equation 5.102

$$\frac{Dm_i}{Dt} = 2\frac{MW_j}{\Re T_i}V_i \sum_{j \in G} \frac{m_j}{\rho_j}\bar{D}\frac{p_i - p_j}{r_{ij}}\frac{dW_{ij}}{dr}. \tag{5.103}$$

Neighbouring liquid particles need to be left out in the summation, as diffusive transport occurs in interaction with the gas phase. It needs to be mentioned, that the Stefan flow (Baehr and Stephan 2010, p. 82) is not considered in this approach and mass transfer therefore underestimated.

As the gas density is three orders of magnitude lower than the liquid one, direct doupling of the equations of motion is challenging, if physically true values are applied. This will not be treated further, as calculation of gas motion is of major interest when surface tension and wetting effects shall be calculated with the CSF approach rather than by pairwise forces. If only diffusion in the gas phase is to be considered, implementation as an underlying grid is much more efficient.

Diffusion Driven Drying Employing Particle Grid Coupling

A full consideration of the gas just for sake of calculating diffusive transport is very inefficient. As an alternative, the gas can be represented by an underlying grid, in which the single nodes are activated depending on their coverage by SPH particles. The easiest grid configuration consists of a regular, quadratic/cubic node distribution with a mesh size of the particle spacing l_0. A natural way of evaluating a grid node's coverage consists in an SPH summation

$$\varphi_{i \in Grid}^{SPH} = \sum_{j \in SPH} \frac{m_j}{\rho_j}W_{ij}, \tag{5.104}$$

in which i is a grid node and j runs over all SPH particles within the vicinity of i. Grid points having a volume fraction φ_i^{SPH} of SPH particles lower than a certain threshold are active. A sensible limit value is 0.5.

Alternatively, the coverage can be calculated in treating both kinds of computational nodes as geometrical objects, for instance squares/cubes or circles/spheres. The intersecting volumes of a gas node with SPH particles in its vicinity are summed up and compared with the volume of the gas point, which again yields a volume fraction φ_i^{geo}. The first approach has in advantage, that it is motivated by an SPH summation, which is consistent with the particle field. The second method is computationally more efficient, as only very near SPH particles contribute (intersecting squares/cubes only share a volume for $r_{ij} < l_0\sqrt{d}$) and unnecessary square-root and kernel evaluations are avoided. As coverage is only interesting near the liquid-air interface, the numerical procedure could be sped up by various advanced algorithms, though. Simulations in this work employed the SPH summation for consistency reasons.

The coupling between SPH particles and neighbouring grid nodes at the liquid-gas interphase can be simply achieved by applying equation 5.103 to the liquid surface particles, with the grid nodes acting as virtual SPH neighbours j. In the same way the contribution of interfacial mass flux to the gas nodes is to be considered by equation 5.102 with neighbouring particles $j \in SPH$. The grid is therefore treated as stationary SPH particles. In principle, the diffusion equation 5.102 could be used in order to calculate diffusive vapour transport within the gas phase. As gas nodes are aligned on a regular grid, it is however more efficient to apply a simple finite volume approximation. Considering an active control volume i in quadratic/cubic grid configuration, the vapour flux at the interface Γ to a neighbouring active nodes j is

$$\vec{J}_v^\Gamma = -D\frac{p_{v,j} - p_{v,i}}{l_0}\vec{e}_{ij}, \qquad (5.105)$$

when the mesh size is l_0. The change in vapour pressure due to diffusion on the grid is hence

$$\frac{dp_{v,i}}{dt} = \sum_j D\frac{p_{v,j} - p_{v,i}}{l_0^2}, \qquad (5.106)$$

with j denoting adjacent active volumes. This is just a centred difference operator leaving out the contribution of inactive gas nodes. The contribution of interfacial mass transfer by equation 5.102 is then simply added as a source term.

Possible Further Model Development

In the present implementation, heat transfer is not regarded. Addition of a temperature equation to the gas grid is straightforward. Implementation of the Stefan flow would yield a more realistic diffusion behaviour. The approach of the gas grid could be applied to droplet drying by calculation of the gas flow around the droplet with an appropriate method. This would allow for individual heat and mass transfer and spatially inhomogeneous drying regimes at different droplet locations. Alternatively, the analytical solution of a potential flow around the droplet could be imposed to account for the flow field in the gas.

5.8 Time Integration

Smoothed Particle Hydrodynamics is a spatial discretisation method. Integration with respect to time can be done by established methods, but needs to account for SPH peculiarities. Compared to grid-based methods, the discretisation stamp is very large. If a neighbourhood radius of three times the particle size l_0 is used - a lower boundary in many applications -, about 30 particles ($\pi 3^2$) are contained in a two-dimensional discretisation and more than 110 ($\frac{4}{3}\pi 3^3$) in three dimensions. Evaluation of single terms is therefore very expensive. Particle motion continuously changes the discretisation stamp in terms of the neighbouring particles to be included and the relations W_{ij} and $\nabla_i W_{ij}$. The Jacobi matrix of an implicit solver would therefore become very complicated, which makes a fully implicit integration practically impossible. For the sake of numerical efficiency, lower-order methods are often preferred (Rosswog 2009) in order to avoid multiple function evaluations during a time step. Traditional SPH solvers are leap frog/Verlet or predictor-corrector schemes. The typical ISPH scheme (section 5.4) employs a predictor-corrector approach as well, but only provides first order accuracy (Cummins and Rudman 1999) because an explicit treatment of accelerations in the predictor step is used. Bøckmann, Shipilova, and Skeie (2012) introduce an ISPH integration based on a BDF-2 scheme with additional function evaluations at half-steps. Their approach exhibits superior accuracy compared to a WCSPH variant, but a comparison with traditional ISPH had not been undertaken and it remains unclear whether the additional solution of the pressure Poisson equation is of great advantage. Pressure solution is by far the most costly part of ISPH so that the traditional scheme as depicted in section 5.4

was applied to all calculations in this work.

If equations do not involve particle motion, an implicit solution is possible. This can be advisable when different physical effects may be decoupled. An equation with a very strict stability criterion for explicit integration either needs to be calculated in a subloop or can be integrated employing much larger time steps in an implicit way, if the overall solution shall not be slowed down by reducing the general time step size. If the effect of particle displacement is small compared to the other terms in an equation, the implicit time step size can be chosen (to some degree) independent from particle motion time stepping. This is applied to the energy balance in the current model, which is solved implicitly in substantially larger time steps than the equations of motion.

5.8.1 Stability Criteria in Explicit Time Stepping

When integration is undertaken by explitid schemes, stability criteria need to be satisfied depending on the implemented physical effects. Whereas in a grid-based method an information should only be moved by a fraction of the distance between nodes/the cell width, SPH particles should only advance by a fraction of the smoothing length.

Convective Transport

Time stepping concerning particle motion is constrained by a typical CFL criterion (Courant, Friedrichs, and Lewy 1928, Courant-Friedrichs-Levy). The time step size in the ISPH method is calculated by (Cummins and Rudman 1999; Shao and Lo 2003)

$$\Delta t_v \leq \beta_v \frac{h}{v_{max}}. \tag{5.107}$$

The limiting prefactor β_v is chosen to 0.25 by Cummins and Rudman, whereas Shao and Lo use a value of 0.1. The weakly compressible SPH employs an equation of state, which additionally involves the propagation of sound waves. The CFL criterion therefore needs to be based on the speed of sound c, which typically is ten times larger than the maximum velocity in order to limit density fluctuations

$$\Delta t_c \leq \beta_c \frac{h}{c}. \tag{5.108}$$

If an artificial viscosity is employed for the stabilisation of WCSPH, this needs to be taken into account for the maximum time step size (Lattanzio et al. 1985;

Monaghan 1992), see below for Δt limitation with respect to viscous flow. These CFL equations are overrestrictive in case of uniformly moving particle collectives. It can therefore be sensible to employ the maximum velocity of particles relative to their neighbours instead of the simple maximum of all velocities. Still, the ISPH literature typically refers to v_{max}, whereas the WCSPH method anyway necessitates the speed of sound, which is much higher than v_{max}.

Acceleration and External Forces

Condensing all right hand side terms of the momentum balance equation to $\frac{D\vec{v}}{Dt} = \vec{a}$ allows for a stability criterion

$$\Delta t_a \leq \beta_a \sqrt{\frac{h}{a_{max}}}, \qquad (5.109)$$

which accounts for momentum change based on the maximum particle acceleration (Lattanzio et al. 1985; Monaghan 1992; Morris, Fox, and Zhu 1997). β_a is typically chosen to 0.25. Surface tension forces, if calculated by the CSF approach, additionally involve the propagation of capillary waves and require a respective stability criterion. Further information on this topic can be found for SPH in Morris (2000) and in general in the original CSF publication from Brackbill, Kothe, and Zemach (1992). A corresponding criterion for the implementation using interparticle forces has not been discussed in the literature to the author's knowledge. The contribution of these forces is therefore considered by the criterion for the acceleration in the present model. Yet, interparticle forces can exhibit a behaviour similar to springs so that oscillations and their propagation are possible in principle. Future research might approach this point.

Diffusive Transport

Diffusive transport involves heat conduction, matter diffusion and viscous momentum transport. These effects are of similar form

$$\frac{Df}{Dt} = \nabla K \nabla f$$

and therefore use the same type of stability criterion ($K \left[\frac{m^2}{s} \right]$ being a transport coefficient). Morris, Fox, and Zhu (1997) adapt typical finite differences criteria

215

in order to derive an SPH analogon for viscous transport within a Newtonian medium by replacement of the node distance with the smoothing length to

$$\Delta t_\eta \leq \beta_\eta \frac{\rho h^2}{\eta},$$ (5.110)

with $\beta_\eta = 0.125$. Zhu and Fox (2001) rewrite this formula for matter diffusion

$$\Delta t_D \leq \beta_D \frac{h^2}{D},$$ (5.111)

again with $\beta_D = 0.125$. The maximum time step size for heat conduction is analogously (Cleary 1998; Cleary and Monaghan 1999)

$$\Delta t_\lambda \leq \beta_\lambda \frac{\rho c_v h^2}{\lambda}.$$ (5.112)

Cleary (1998) and Cleary and Monaghan (1999) set β_λ between 0.1 and 0.15.

In any case these criteria can become very restrictive if the smoothing length / particle spacing becomes very small (or the transport coefficient is high). This can be a severe problem when refining the resolution. When diffusive transport is limiting, doubling the resolution by taking half of the smoothing length involves four or eight times the particles in 2D and 3D, respectively, and four times as many time steps. One advantage of ISPH - that its minimum time step size is one order of magnitude higher compared to WCSPH - vanishes, if the viscous criterion limits the time steps to the magnitude of the WCSPH CFL criterion. This "curse of a fine resolution" can only overcome by implicit calculation of the respective physical effects.

5.8.2 Time Stepping Criteria Employed in This Work and Their Reference Length

As described in section 5.1.6, the definition of h is to some extend arbitrary, which also affects the definition of stability criteria. In engineering applications of incompressible liquids the fluid is represented by particles of constant size l_0, which are densely packed and fairly homogeneously distributed (at least this is the ideal representation). It appears reasonable to restrict time stepping such that particles move only part of an average particle distance (l_0) rather than in comparison to the smoothing length, which depends on the choice of the kernel and the cut-off radius. If the quintic spline ($h = 1/3r_{cut}$) and the cubic spline kernel ($h = 1/2r_{cut}$) were used at the same cut-off radius, the latter would allow for a 50 % higher time step size. Particles being exposed to the same attractive forces would therefore be allowed to approach each other much farther in case of the cubic spline, which does not appear reasonable concerning numerical stability. This was also discussed shortly by Morris, Fox, and Zhu (1997), who attributed the various splines a different "effective" resolution length for the same value of h and suggested a modification of CFL coefficients in case of the cubic spline. As already stated before, one of SPH's deficiencies is the lack of a comprehensive mathematical foundation and analysis, especially in case of disordered particles, so that time stepping criteria are often chosen according to numerical experiments and common experience.

In order to circumvent the uncertainties concerning the definition of h and because the particle spacing appears a reasonable reference value, time stepping criteria in this work have been based on l_0. This might be over-restrictive, because the smoothing length should generally be chosen larger than the particle spacing. As the quintic spline in combination with a cut-off radius of $3.1l_0$ was applied throughout all simulations, the difference between smoothing length and particle spacing values is not large, though, and both values nearly coincide.

The particle motion criterion 5.107 and the acceleration restriction 5.109 are used in the present model to limit the time step size. The diffusive criteria would restrict the calculation such that an integration of the problem in a feasible amount of time would scarcely be possible, especially when refining the resolution. Heat conduction was therefore implemented in an implicit way with much larger time steps than the equations of motion (see section 5.8.3 for details). The viscous criterion was dropped without problems of numerical instability. A further analysis has not been carried out. Possible reasons may be that the semi-implicit nature of the ISPH method improves the numerical stability, that the superposition of interparticle forces dampens viscous oscillations or that the presence of rigid bodies effectively limits the propagation of perturbations.

5.8.3 Implicit Solution of Diffusive Equations

Matter transport involving Fickian diffusion is expressed in Lagrangian notation as follows:

$$\frac{D\rho_j}{Dt} = -\nabla\left(\rho_j \vec{v}\right) + \nabla\left(D\nabla\rho_j\right) + r_j^F MW_j + \vec{v}\nabla\rho_j \qquad (5.113)$$

$$\frac{D\rho_j}{Dt} = -\rho_j\nabla\vec{v} + \nabla\left(D\nabla\rho_j\right) + r_j^F MW_j = \nabla\left(D\nabla\rho_j\right) + r_j^F MW_j. \qquad (5.114)$$

After application of the divergence free condition due to incompressibility only the diffusive and source terms remain. The same holds for the energy balance in form of a temperature equation

$$\frac{DT}{Dt} = \frac{1}{\rho c_p}\nabla\left(\lambda\nabla T\right) + \left.\frac{DT}{Dt}\right|_{LDF} + \left.\frac{DT}{Dt}\right|_{evap}. \qquad (5.115)$$

The last two terms condense the contribution of linear driving force heat transfer and cooling by evaporation as described in section 5.7.2. The following, simple implicit approaches for the temperature equation (Euler, trapezoidal rule / Crank-Nicholson and BDF-2)

$$T_i^{n+1} - \Delta t\frac{DT_i^{n+1}}{Dt} = T_i^n \qquad (5.116)$$

$$T_i^{n+1} - 0.5\Delta t\frac{DT_i^{n+1}}{Dt} = T_i^n + 0.5\Delta t\frac{DT_i^n}{Dt} \qquad (5.117)$$

$$3T_i^{n+1} - 2\Delta t\frac{DT_i^{n+1}}{Dt} = 4T_i^n - T_i^{n-1}, \qquad (5.118)$$

can be expressed in a general way as

$$f_v^{n+1}T_i^{n+1} + f_t^{n+1}\Delta t\frac{DT_i^{n+1}}{Dt} = f_v^n T_i^n + f_v^{n-1}T_i^{n-1} + f_t^n\Delta t\frac{DT_i^n}{Dt}. \qquad (5.119)$$

The left hand sides of the equations contain the unknown values at time step $n+1$, whereas the values on the right hand side are known.

The prefactor f_v^{n+1} and f_t^{n+1} depend on the desired method (see Table 5.1). Equation 5.119 can be rewritten in matrix vector notation as

$$A\mathbf{T} = \mathbf{b}, \qquad (5.120)$$

in which all unknown particles' temperatures are condensed in the vector $\mathbf{T}$ and $\mathbf{b}$ is the constant right hand side vector. The matrix A is determined by the discretisation of the heat conduction equation and the contribution of the respective

Table 5.1: Prefactors for the respective method within the generalised implicit scheme

method	f_t^n	f_t^{n+1}	f_v^{n+1}	f_v^n	f_v^{n-1}
implicit Euler	0	-1	1	1	0
BDF-2	0	-2	3	4	-1
trapezoidal rule	0.5	-0.5	1	1	0

particles to heat transfer. In a wholly implicit scheme cooling by evaporation would have to be calculated with respect to the current temperature T^{n+1}. Due to the non-linear dependency of the vapour pressure on the temperature, the problem would become non-linear as well and the solution much more elaborate. Hence, the cooling by evaporation is calculated based upon the temperature of the previous step T^n and inserted into the constant right hand side vector. The error is small, though, and the system is self-stabilising. If the temperature is too high, evaporation will be stronger and thus limit the future temperature and vice versa. This way the average temperature and cooling rates are calculated correctly.

The detailed matrix and vector formulae are

$$A_{ii} = f_v^{n+1} + \frac{f_t^{n+1}}{\rho_i c_{p,i}} \left(\sum_j 2 \frac{m_j}{\rho_j} \bar{\lambda}_{ij} \frac{1}{r_{ij}} \frac{dW_{ij}}{dr} - \left| \frac{\nabla c_i}{[c]} \right| \alpha \right) \tag{5.121}$$

$$A_{ij} = -\frac{f_t^{n+1}}{\rho_i c_{p,i}} 2 \frac{m_j}{\rho_j} \bar{\lambda}_{ij} \frac{1}{r_{ij}} \frac{dW_{ij}}{dr} \tag{5.122}$$

$$b_i = f_v^n T_i^n + f_v^{n-1} T_i^{n-1} + f_t^n \frac{\partial T_i^n}{\partial t} - \frac{f_t^{n+1}}{\rho_i c_{p,i}} \left(\left| \frac{\nabla c_i}{[c]} \right| \alpha T^\infty + \frac{\Delta h_v}{V_i} \frac{dm_i^{n+1}}{dt} \right). \tag{5.123}$$

The system of linear equations can be solved by the same kinds of numerical methods as for the PPE solution in the ISPH method. As the matrix A is non-symmetric, the conjugate-gradient method cannot be applied, but other solvers of the Krylov subspace family as GMRES or BiCGStab are well-suited (Meister and Vömel 2008). The calculations throughout this work have been conducted using the GMRES solver from the PETSc package (Balay et al. 2019). Numerical tests have been performed using all three implicit methods mentioned above. Productive simulations of drying were performed using the trapezoidal rule.

5.8.4 Initialisation of an SPH Calculation

As SPH is integrated using explicit methods, the initial particle and value configuration affects the outcome of the simulation and has to be carefully chosen. This is especially true for the incompressible WCSPH method, which involves oscillations of the pressure and velocity fields. Examples for initialisation procedures in WCSPH can for instance be found at Monaghan (1994) and Tartakovsky and Meakin (2005). In comparison, the ISPH method is significantly more robust, as the calculated pressure field is already near to the correct solution within the first time step and converges throughout the following steps. Still, the initial particle configuration will have an effect on boundaries and interfaces, which can exhibit steps depending on the particle configuration. The overall effect of a stepwise boundary discretisation will be diminished with a finer resolution. Typically, particles are initially not distributed randomly throughout the domain, but in a regular alignment. The most simple configuration is a quadratic or - in 3D - cubic particle distribution. Considering the regularity and the number of nearest neighbours, a hexagonal or face-centred cubic alignment are preferable. Setting the particles on the nodes of a regular triangular mesh for a 2D simulation, each particle will have six nearest neighbours in a hexagonal arrangement in comparison to four neighbours in a quadratic alignment. This also improves the representation of curved interfaces. By way of contrast, a consistent initialisation involving periodic boundary conditions is simpler using a quadratic/cubic alignment due to its flat surfaces. Typically, particles are loaded and the particle volume V_i is determined by l_0^d and the particle mass m_i calculated by $V_i \rho_i^0$. If the continuity equation is treated as an initial value problem, the particle density ρ_i is set according to the physical density ρ_i^0 as well. Otherwise, and typically in ISPH, it is determined throught each simulation step by summation (equation 5.32).

Drying calculations employed in this work do not necessitate a certain initialisation procedure. The ISPH solver in combination with rigid bodies and surface tension forces relaxes quickly to a quasi steady state. The temperature evaluation is done fully implicit and not prone to stability issues. Hence, a relaxation procedure was not necessary. Calculation examples with the surface tension approach may need relaxation to a circular drop first.

6. VALIDATION OF THE SPH IMPLEMENTATION

The basic concept of ISPH free surface flow solvers in combination with a Newtonian liquid has been widely applied. Despite some changes, the implementation is quite near to literature approaches. Therefore, not all standard test cases will be stressed in the following. Special regard will be paid to the heat and mass transfer by linear driving forces, as this is a new topic to SPH. Furthermore, the flow solver will be tested with test cases on incompressibility solution. The surface tension and wetting behaviour will be compared for both the isotropic pairwise force and the surface-lateral forces approach.

6.1 Implicit Solution of Heat Conduction

A test case of Cleary and Monaghan (1999) consists of a domain of length L with a sinusoidal temperature profile

$$T(x, t = 0) = \sin \frac{\pi x}{L}. \tag{6.1}$$

At the boundaries $x = 0$ and $x = L$ the temperature is kept constant throughout the computation. The analytical solution for this problem is

$$T(x,t) = T(x, t = 0) \cdot exp\left(-\frac{\lambda \pi^2}{\rho c_p L^2} t\right). \tag{6.2}$$

The SPH discretisation was performed two-dimensionally as in Cleary's and Monaghan's work (1999) using a regular, Cartesian alignment. The problem length L was set to 1 m, corresponding with 200 particles in x-direction for a particle size $l_0 = 0.005\,\mathrm{m}$. In y-direction just 10 particle layers were used in conjunction with periodic boundary conditions. The thermal conductivity was set to $\lambda = 1\,\mathrm{W/(m\,K)}$, the heat capacity to $c_p = 1\,\mathrm{J/(kg\,K)}$ and the density to $\rho = 1\,\mathrm{kg/m^3}$.

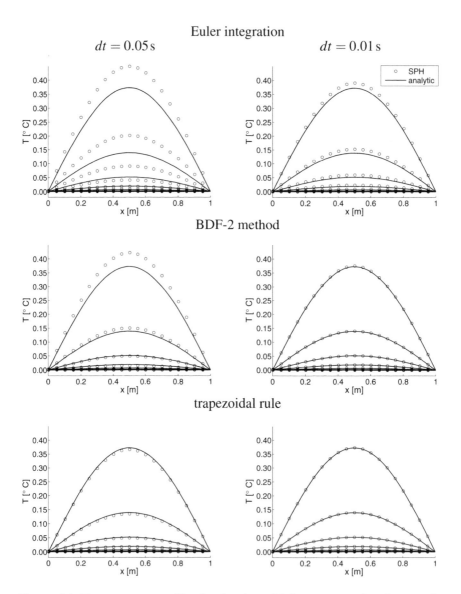

Figure 6.1: Temperature profiles for the sinusoidal test case at time instants from 0.1 s to 1 s in intervals of 0.1 s for the case of mirror boundaries. Time step sizes were 0.05 s (left column) and 0.01 s (right column).

Profiles over time are depicted in Figure 6.1. With a coarse time stepping of 0.05 s, Euler and BDF-2 solvers exhibit large errors, whereas the trapezoidal rule is fairly accurate. At the lower time step size of 0.01 s the results of the two second order methods converge quickly. The Euler solution still displays visible errors. The overall convergence behaviour is shown in Figure 6.2. The convergence over time (left frame) is as expected of only first order for the Euler method, whereas BDF-2 and the trapezoidal rule provide second order convergence.

Convergence with decreasing particle size is depicted in the right frame for two different implementations of the Dirichlet condition. The naive implementation of appointing fixed temperature values to a layer of boundary particles (dotted lines) deteriorates the smoothness of the temperature field and hence provides only first order convergence. If the particle value is mirrored at the boundary, the SPH approximation yields a convergence rate of second order. This example underlines that boundary treatment is a crucial point in SPH. If no proper boundary treatment is possible, the method suffers from its approximation deficiency there.

In many cases the error approached a distinct minimum, with a locally very high convergence rate. Further decrease of time step or particle size does not reduce the error further but rather approaches a limit error above the observed minimum, which cannot be undercut by a finer resolution.

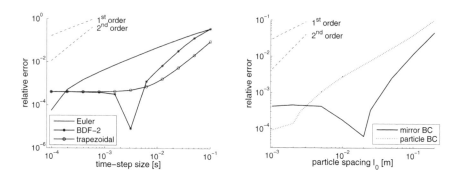

Figure 6.2: Convergence behaviour of the implicit SPH heat conductivity solution: Error at $x = L/2$ and $t = 0.1$ s. Convergence concerning time (left) and spatial resolution (right, only trapezoidal rule).

6.2 Linear Driving Force Heat and Mass Transfer

6.2.1 Heat Transfer to a Unilaterally Heated Rod

Heat conduction in a rod clamped to a hot wall and cooled by the surrounding medium is a classic example in heat and mass transfer. The problem is depicted in Figure 6.3. Initially, a rod of length L has a constant temperature T_0 anywhere. At $t = 0$ the wall is abruptly raised to the temperature T_w. The surrounding gas temperature T^∞ equals T_0. Heat transport occurs inside the rod by conduction and between the rod and the surrounding gas by conduction and convection subsumed in the heat transfer coefficient α (equation 2.122). The steady state solution for $t \to \infty$ is (Baehr and Stephan 2010, p. 125f) with $k^* = \sqrt{\frac{2\alpha}{\lambda R_{rod}}}$

$$T(x) = T^\infty + (T_w - T^\infty) \frac{\cosh\left[k^*(L-x)\right] + \frac{\alpha}{k^*\lambda} \sinh\left[k^*(L-x)\right]}{\cosh(k^*L) + \frac{\alpha}{k^*\lambda} \sinh(k^*L)}. \qquad (6.3)$$

Three-dimensional SPH simulations of this problem have been carried out for a rod of a length $L = 0.25\,m$ and diameters D of 0.05 and 0.10 m. T_w was set to $1\,°C$, whereas $T_0 = T^\infty = 0\,°C$. The heat conduction coefficient in the surrounding gas was $\lambda_G = 0.025\,Wm^{-1}K^{-1}$. Other material properties were unity, as in the test case before. The rods were discretised with particles of $l_0 = 2.5 \times 10^{-3}\,m$. Time stepping was $\Delta t = 0.1 \times 10^{-3}\,s$. The left boundary was implemented by mirroring, the superior approach in the previous example. Mass transfer was modelled according to equation 5.96 in the implicit solution by the trapezoidal rule.

Profiles over time are compared to a simple one-dimensional finite differences solution in Figure 6.4. SPH solutions match the FDM references very well for different rod diameters and Nusselt numbers. The temperature at the rod's end over time (Figure 6.5) is also very close to the FDM solution.

The CSF implementation of heat transfer applies the boundary condition in the desired manner and can be further used for drying applications.

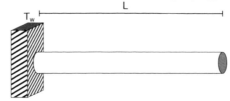

Figure 6.3: Geometry of a unilaterally heated rod.

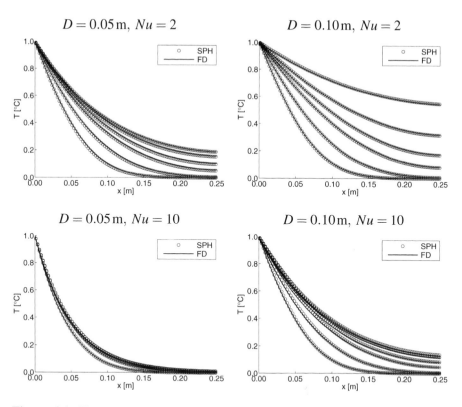

Figure 6.4: Temperature profiles within a unilaterally heated rod for different Nusselt numbers and diameters (2, 4, 8, 12, 20, 200 s = steady state).

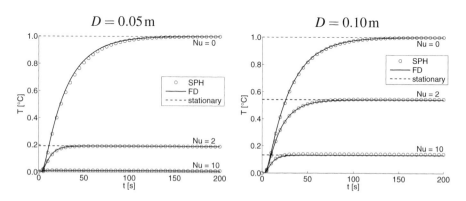

Figure 6.5: Temperature at the rod's end over time for various Nusselt numbers.

6.2.2 Coupled Heat and Mass Transfer: Droplet Evaporation

Evaporation of a pure water drop in hot air is a problem of coupled heat and mass transfer, which efficiently can be solved in a lumped manner with the equations derived in chapter 3. Material parameters and initial and boundary conditions are provided in Table 6.1. Nusselt and Sherwood numbers were both set to 10. Simulations have again been carried out in three dimensions. Heat and mass transfer between droplet and gas were implemented according to section 5.7.2, heat conduction within the droplet corresponding with equation 5.87.

The course of the droplet radius over time is drawn for different particle sizes in Figure 6.6. Results from the fine resolution match the 0D reference, whereas the coarse resolution does not resolve the drying behaviour well. The radius according to the outmost particle's position ("SPH, part") exhibits a step-wise drying behaviour. The size of an SPH particle cannot be altered easily (section 5.7.4). While the evaporative mass is reduced, the ordinary mass and, thus, the outer droplet shape remain unaltered until the SPH particle is deleted. Then, suddenly, the droplet becomes smaller. By this, the surface area is over-estimated, which explains why evaporation is too intense in the coarse example. A better way of evaluating the droplet radius considers all SPH particles' evaporative mass as belonging to a perfectly shaped drop ("SPH, mass" in the graphs)

$$R_{mass} = \left(\frac{3 \sum m_i^{evap}}{4\pi\rho_l} \right)^{1/3}. \tag{6.4}$$

In doing so, the radius shrinks continuously and sub-surface particles are included. This value is not only smooth (the solid line with dots), but matches the reference significantly better. The right frame of Figure 6.7 shows the sur-

Table 6.1: Material properties and conditions for droplet evaporating.

material properties			initial and boundary conditions		
Δh_v	-2.257×10^6	J/kg	T_0	20	°C
ρ^L	1000	kg/m^3	T^∞	180	°C
c_p^L	4200	J/kg	p_v^∞	0	Pa
λ^L	0.6	W/(m K)			
λ^G	0.025	W/(m K)			
D^G	30×10^{-6}	m^2/s			

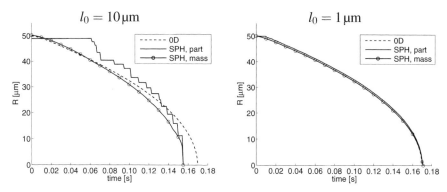

Figure 6.6: Droplet drying calculated by SPH: Radii over time from the outmost particle's position ("part") or an equivalent sphere of the same mass ("mass").

face area calculated via the CSF approach in the SPH solution. It was obtained by summing up all interface area to volume ratios according to equation 5.95 multiplied with the particle volume V_i. The CSF approximation is resolution dependent. Using a larger number of particles, the droplet surface is less step-wise and rather smooth and its area is not overestimated any more.

The droplet temperature over time is shown in Figure 6.7, left frame. SPH and the 0D reference reach the same wet bulb temperature. Errors in the SPH linear driving force calculation affect heat and mass the same way so that deviations cancel out in the balance of both effects. Due to perfect mixing and intensified heat transfer, thermal equilibration is slightly faster in the 0D model. The SPH result shows the same course over time, otherwise.

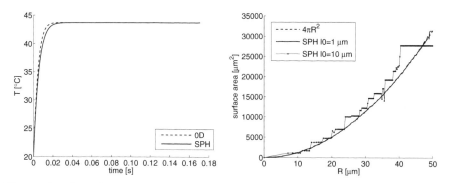

Figure 6.7: Droplet temperature over time (left) and SPH/CSF surface area vs. droplet radius (right).

6.3 Diffusion Driven Drying by SPH-Grid Coupling

The following example considers a stagnant water column in an infinitely stretched domain, which is covered by a layer of air. At a certain, fixed height h_0 the vapour pressure is fixed to $p_v = 0$. At the surface of the water column the air is saturated with $p_v = p_v^S$. Assuming an isothermal setup and neglecting the Stefan blow, the height h of the water column will evolve according to

$$\frac{dh}{dt} = -\frac{MW_{H_2O}}{\rho_{H_2O}^0} \frac{D_G}{\Re T_G} \frac{p_v^S - p_v(h_0)}{h_0 - h}. \tag{6.5}$$

This problem can be treated by the approach depicted in section 5.7.5. The liquid is represented as stationary SPH particles. An underlying grid is used to discretise the gas in regions uncovered by SPH. Coupling of mass transfer between grid and SPH particles is performed by equation 5.102. Mass loss of SPH particles is calculated according to equation 5.103. Physical constants are summed up in Table 6.2. Example calculations have been carried out for a water layer of initial height $h = 100\,\mu m$ and a fixed vapour pressure of $0\,Pa$ at $h_0 = 215\,\mu m$. For reasons of simplicity, the grid discretisation was performed with the same meshsize as the SPH particle spacing on a quadratic grid.

Figure 6.8 shows the setup on the left. Particles are depicted as blue circles. The gas layer is coloured according to the vapour pressure. The graphs on the right show the evolution of the water column height over time for three different particle spacings l_0. In order to avoid a stepwise profile over time, the water level has been evaluated according to

$$h_{SPH}^t = \max(y_{SPH}) - \min(y_{SPH}) + \frac{m_{i_{Mx}}^{evap}}{m_{i_{Mx}}} \cdot l_0. \tag{6.6}$$

Table 6.2: Material properties and conditions for the evaporating water column.

liquid properties		
Δh_v	-2.257×10^6	Jkg^{-1}
ρ^L	1000	kg/m^3
c_p^L	4200	$J/(kg\,K)$
T^L	60	$°C$
D^G	30×10^{-6}	m^2/s
$p_v(h_0)$	0	Pa

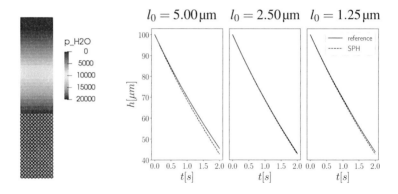

Figure 6.8: Calculation of diffusion driven drying of a water column: liquid particle and gas grid setup (left), evolution of water column height over time for different resolutions (right).

i_{Mx} denotes the particle at the highest position in the water column. By taking the mass loss within the upper particle layer into account, the curve is smoothed. The overall results represent the correct physical behaviour and the SPH results follow the reference line obtained by numerical solution of equation 6.5 with a standard DoE solver[1]. The coarse discretisation shows the greatest deviation (left). Yet, the convergence behaviour is not fully clear, as the smallest particle size shows a slightly greater deviation from the reference than the particle size in between. Time integration was performed by the explicit Euler method for solution of the diffusion equation. As diffusion driven drying was not the main focus of this work, implementation of a higher order scheme was not undertaken and the diffusion equation was solved in a subloop at small time step sizes ($\Delta t \leq 0.2\frac{l_0^2}{D}$) due to the explicit method. Implementation of a higher order implicit scheme, as it is used for heat conduction, is surely advisable, when the coupled approach of diffusion driven drying shall be used further. Still, the numerical results are satisfactory when compared to the reference.

---

[1]Solution obtained with the standard explicit Runge-Kutta method of the Scipy package.

6.4 SPH Flow Solver

In the following, the underlying flow solver used in the drying model is validated by several test cases. As already stated, numerous validation cases can be found in the literature (e.g. Colagrossi 2005; Keller 2015; Monaghan 1994) and not all of them will be presented here as the current code is very near to a standard ISPH implementation. All further tests have been conducted in two dimensions.

6.4.1 ISPH Solution of a Standing Water Column

The pressure profile of a stagnant water column is shown in Figure 6.9. The height of the water column is 1 m. With a density of $1000\,\mathrm{kg/m^3}$ and gravitational acceleration of $9.81\,\mathrm{m/s^2}$ the maximum pressure is $\rho gh = 9810\,\mathrm{Pa}$. The profile was obtained after 100 time steps of 10×10^{-4} s. The SPH solution matches this theoretical value and the linear profile very well. The second point above the ground exhibits some deviation, which is caused by the boundary condition. This implementation used the fairly simple approach of Shao and Lo (2003), who just applied a Dirichlet condition by taking the upper boundary row into the pressure Poisson matrix and copying this values to particles being further away from the water column (see section 5.4.1). Hence, the pressure profile is not continuous over the boundary, which slightly affects liquid particles nearby. Concerning free surface flow in a droplet, this peculiarity of a certain boundary condition is not relevant.

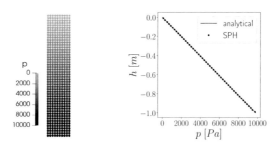

Figure 6.9: Simulation of a stagnant water column, left: particle setup, right: pressure profile.

6.4.2 Free Surface Flow

A circular drop of incompressible liquid will be deformed by a velocity field

$$v_x(x,y) = -A_0 x \tag{6.7}$$

$$v_y(x,y) = A_0 y, \tag{6.8}$$

to an ellipse (Monaghan 1994; Colagrossi 2005; Keller 2015). With the function

$$A(t) = -\frac{\dot{a}}{a} = \frac{\dot{b}}{b}, \tag{6.9}$$

the evolution of the ellipse's semiaxes a and b is known. A is defined as (Keller 2015)

$$\frac{dA}{dt} = \frac{A^2\left(a^4 - a^2 b^2\right)}{a^4 + a^2 b^2}. \tag{6.10}$$

By this, the evolution of the ellipsis over time can be solved numerically as a system of ordinary equations. The pressure within the drop is expressed as

$$p(x,y,t) = \frac{\rho}{2}\left(\dot{A}\left(x^2 - y^2\right) - \dot{A}\left(x^2 + y^2\right) - a^2\left(\dot{A} - A^2\right)\right). \tag{6.11}$$

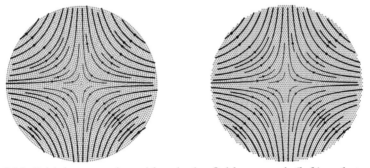

Figure 6.10: Initial geometries with velocity fields: smooth (left) and step-wise (right) drop surface, line widths indicate the velocity magnitude.

Figure 6.10 shows the initial geometries with the initial flow field, in the left frame with a circular particle alignment and a smooth surface, in the right frame the hexagonal case with a more irregular surface, in both cases with $l_0 = 0.025$ m. The droplet radius was set to 1 m, the density to $1000\,\text{kg/m}^3$ and the initial value $A_0 = 1/\text{s}$. The time step size was 0.01 s. Figure 6.11 contains

calculations with both initial geometries. Two implementations of the free surface condition have been tested. In the literature approach a density threshold of 97 % was used in order to impose the boundary condition according to Keller (2015) and Bøckmann, Shipilova, and Skeie (2012) within the PPE (see section 5.4.3). The penalty approach does not use separate surface particle testing and enforces a penalty function depending on the $\nabla \vec{x}$ value as described in section 5.4.4. The final geometries after streching the drop are shown on the left. In case of the very regular surface both approaches yield a virtually identical final profile. Differences within the surface are comparably small so that the effect of free surface switching becomes negligible. The evolution of the half-axes and the pressure over time are drawn in the upper right graphs. The analytical solution is matched very well concerning the geometry and for the most part for the pressure. Initial pressure build-up is somewhat retarded, which may be due to the geometry which is optimised for a smooth surface, but exhibits particle disorder and a high density in the centre. Otherwise the course of the pressure over time is reproduced very well with some small oscillations, which have also been reported by Keller (2015).

Simulations of the more step-wise geometry exhibit visible deviations, first for the pressure field, then - as a consequence - in the evolution of the semi-minor axis a. This affects both implementations of the boundary condition, but the penalty based approach keeps closer to the theoretical pressure so that the final elliptic shape is matched significantly better. The middle picture shows the final geometry in case of the penalty based approach, the right one the literature solution. The penalty condition does not distinguish between free surface and bulk particles in a hard way, but provides a smooth transition, which helps to calculate a more regular pressure and flow field. This is visible in most cases of non-ideal geometries for this test case. Additionally, the literature approach demands zero pressure truncating, as simulations crashed otherwise when negative values occured. Pressures below zero were not a problem when using the penalty based free surface condition, which again indicates that this boundary treatment provides a smoother solution of the pressure field.

The problem of the deforming drop is a special case which is very sensitive to disturbances of the surface and the particle order there. When in other cases a droplet needs to be initialised, the hexagonal alignment is typically a good start and involves only minor re-organisation of particles for instance under surface tension. The smooth droplet geometry performs worse in such cases, as it

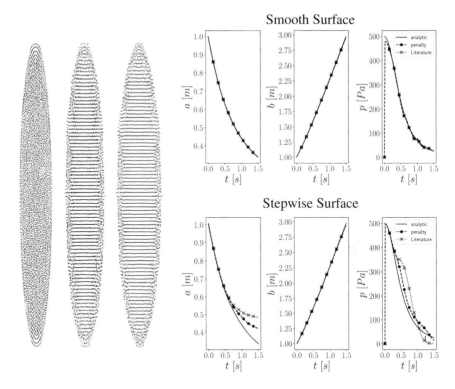

Figure 6.11: Simulation of eliptically deforming drop, on the left final geometries, from left to right: smooth surface (both approaches), rough surface + penalty and + literature approach; on the right: drop deformation and centrical pressure over time for the smooth (upper graph) and rough surface (below).

involves a higher degree of inner particle reorganisation.

6.4.3 Surface Tension Approach of Pairwise Forces

In order to study the surface tension behaviour, a test case of Adami, Hu, and Adams (2010b) was employed. Imposing the flow field

$$v_x = v_0 \frac{x}{R_0} \left(1 - \frac{y^2}{rl_0}\right) \exp\left(-\frac{r}{R_0}\right) \tag{6.12}$$

$$v_y = -v_0 \frac{y}{R_0} \left(1 - \frac{x^2}{rl_0}\right) \exp\left(-\frac{r}{R_0}\right) \tag{6.13}$$

on a circular drop will enforce a deformation, which leads to droplet oscillation. The period of oscillation $\tau = 2\pi\sqrt{\frac{R_0^3 \rho}{6\sigma}}$ depends on the droplet radius, the density and the surface tension coefficient. Figure 6.12 shows the evolution of the droplet shape on the left at different instants of time. The initial radius was 0.4 m and v_0 was set to 5 m/s. Density and particle spacing were 1000 kg/m^3 and 0.02 m, respectively. The interaction parameter s_{ij}^0 in equation 5.75 was set to 1000. The graphs on the right side of Figure 6.12 show the maximum elongation of the droplet from its centre of mass throughout the calculation. The droplet exhibits an oscillating behaviour with decreasing amplitude due to viscous dissipation. If this behaviour was invariant from scaling, the surface tension coefficient could be correlated to the s_{ij}^0 in order to tune this value.

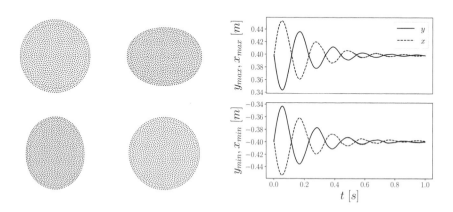

Figure 6.12: Oscillating droplet of 0.4 m radius: The pictures on the left depict droplet shapes at 0, 0.058, 0.173 and 0.230 s. On the right: elongation in y (solid) and x (dashed) direction.

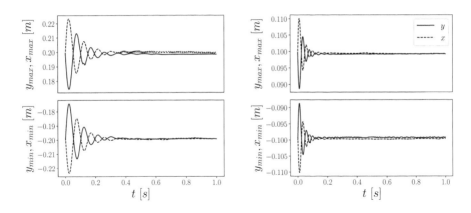

Figure 6.13: Oscillating calculations with radii of 0.2 (left) and 0.1 m (right).

This was the one test case in which the geometry had been relaxed before. A hexagonal drop was exposed to surface tension forces and after equilibration the velocity field was imposed and the calculation restarted. Figure 6.13 shows the oscillations for droplets of the same particle setup, but for radii of 0.2 and 0.1 m. As the relaxed geometry had been reused, the particle sizes l_0 had been adapted accordingly to 0.01 and 0.005 m.

The principle behaviour is according to physics with smaller droplets oscillating faster. The oscillation periods τ had been measured in the postprocessing as the time between every other incident in which x_{max} and y_{max} cross each other. The right frame of Figure 6.14 shows that this value changes over time, especially in case of the larger droplet. $\sqrt{R^3}$ yields values of about 0.03, 0.09 and 0.25 m$^{1.5}$, which are very roughly matched by the oscillation periods in the graph, if $2\pi\sqrt{\frac{\rho}{6\sigma}} = 1\,\mathrm{s/m^{1.5}}$. The scaling of oscillations for different radii is mediocre, as the simulated oscillation periods are closer to each other than implied by their radii. Moreover, the pressure inside the droplet is not scaling properly. According to the Young-Laplace equation, the pressure is proportional to $\frac{1}{R}$. As can be seen in the middle frame, this is not represented well by the model. In particular the small radius yields a pressure which is far too small, but also the ratio between 0.2 and 0.4 m does not match. The evolution of pressures over time is depicted in the left frame (the lowest curve corresponds - wrongly - with the smallest radius, the highest with the radius in between).

Figure 6.15 contains calculations employing the original force from Tar-

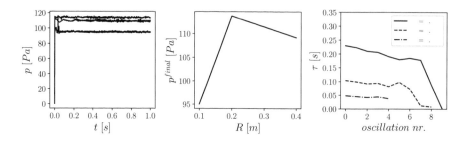

Figure 6.14: Oscillating drops, pressure evolution and oscillation period: Pressure over time (left), steady state pressure (middle) and oscillation period (right) for different initial radii.

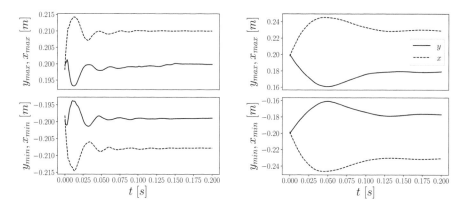

Figure 6.15: Oscillating drops simulated with the isotropic surface tension force of Tartakovsky and Meakin, evolution of elongations over time for different prefactors $s_{ij}^0 = 200$ (left) and 20 (right).

takovsky and Meakin (2005) with s_{ij}^0 being 200 (left graph) and 20 (right). The different parameterisation to the surface-lateral force approach are due to the fact that the latter involves additional multiplications with $1 - \frac{\nabla \vec{x}}{d}$ so that equality of the forces is reached approximately for a ratio of 10 between both. The droplet shows sort of an oscillating behaviour, but does not relax to its initial position.

Finally, the problem of scaling is shown in Figure 6.16 for a droplet of 0.2 m radius, but $l_0 = 0.005$ m being half the size of the previous calculation. Hence, four times the particles were used in this example. The initial velocity profile

Figure 6.16: Oscillating drops, failure of the pairwise force approach: Initial setup (left), surface-lateral force approach and isotropic force according to Tartakovsky and Meakin (right) after 0.08 s.

was the same as before. Now, neither the anisotropic nor the isotropic pairwise force approach were able to stop the drop from elongating.

The pairwise force approach in its current implementation is in best case mediocre in simulating dynamic processes involving large droplet deformations. Moreover, the lack of scaling behaviour is a great deficiency. One reason for this is the lack of orientation. It is challenging to obtain a value, which yields the inplane force behaviour at all (the absolute value of the colour function's gradient/surface normal and the density ratio failed) and still these forces happen between different particle layers due to the discrete nature of the SPH particles' distribution. Moreover, the necessity of keeping particles apart requires a strong repulsive contribution, which is one further weak point of this approach. However, the traditional isotropic approach does not provide any improvement in the current ISPH implementation. Comparison with weakly compressible SPH and an appropriate equation of state would be interesting in order to obtain the contribution which is necessary for interparticle forces to work on an independent scale.

6.4.4 Wetting Phenomena

Whereas the results of the surface-lateral force approach with respect to the dynamic behaviour of surface tension are unsatisfactory, it exhibits an improvement to traditional pairwise forces concerning the static wetting behaviour on the micro scale. The following calculations have been performed for an initially square drop of size $60 \times 60 \mu m$. The dimensions of the droplet have been chosen such, because this is in the same order of magnitude as spray dried droplets. The results can therefore serve as a test of the applicability of the surface-lateral force approach to a droplet drying model. The particle size was $1.5 \mu m$. The interaction coefficient of the force was set to 30 and time stepping size was 1 μs or lower, if required by limit criteria. The other phyiscal values were those of water as in the previous examples. Figure 6.17 shows the evolution of the droplet shape and the wetting behaviour over several instants of time. Whereas the dynamic behaviour is very likely unsatisfactory (consider the previous example of the oscillating drop), the overall behaviour is reasonable and dynamic. The motion of the contact line slows down with the droplet becoming wider, which is partly a physically reasonable behaviour, but may also be attributed to the lack of dynamics in the approach. The last frame does not represent the end of the process. Final droplet shapes for various contact angles are drawn in Figure 6.18. The approach of surface-lateral forces matches the predefined contact angle remarkably well. There is yet a lower limit, below which the wetting behaviour is not further covered. This is due to the discrete nature of the particle approach. It becomes more and more difficult to represent a small angle with discrete particles of a distinct size. From this perspective, this can be considered as a limit of the numerical method and not of the new approach.

Yet, there are some shortcomings. The particle collective does not in every case come to rest. In the test configuration this is the case for a contact angle of 120°. Particles from the droplet's bulk are dragged towards the surface in the

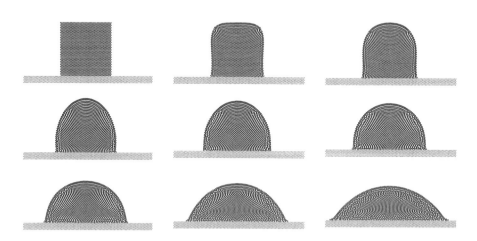

Figure 6.17: Evolution of an initially square drop on a plate, contact angle 30 °, instants of time: 0, 80, 160, 320, 480, 640, 960, 1920, 3840 μs, calculation with the surface-lateral force approach.

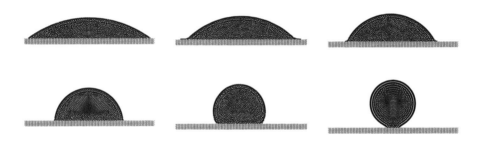

Figure 6.18: Final geometry of sessile drops at contact angles of $0, 30, 60, 90, 120$ and $150°$ calculated with the surface-lateral force approach.

vicinity of the contact line and move upwards in an ongoing swirl. In drying simulations, particles are continuously taken out of the system so that this is a minor issue, but it means that unphysical forces introduce continually kinetic energy into the system. Secondly, the scaling behaviour is again mediocre. The same geometry with a contact angle of $30°$ has been calculated in Figure 6.19 with a smaller and a larger particle size. Both cases do not relax to the final geometry. In case of smaller particles the force particle is scaled down. Concerning the pressure within the droplet, this makes sense as otherwise the force per unit area would grow with decreasing particle size (see the discussion in section 5.5.1). As the neighbourhood in the vicinity of the contact line remains at the same number of particles, the final force is scaled down and too small to drag the liquid bulk further. This issue of scaling has not been solved up to

Figure 6.19: Partial failures of the surface-lateral force approach in reaching a contact angle of $30°$: Same Geometry as before, but with greater $l_0 = 2\,\mu m$ (left) and smaller $l_0 = 1\,\mu m$ (right).

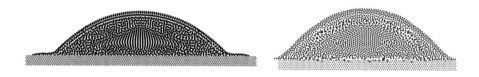

Figure 6.20: Failures of the surface-lateral force approach, contact angle 30°: Calculation without the additional repulsive force (left) and without adding an additional density constraint (right).

now. The example with enlarged particle size exhibits a "snap in" behaviour of the outmost liquid particles into the surface between two solid particles, which prevents further motion of the contact line. This effect is to some degree wanted and caused by the repulsive contribution, but does not scale properly as well. The repulsive force is not only necessary to provide a better particle distribution, but also in order to prevent an ongoing stream of a single particle layer from the contact line along the top of the solid surface. Otherwise, one obtains exactly the opposite behaviour to snapping in as in the left frame of Figure 6.20, in which the additional repulsive force was switched off. Whereas the droplet bulk exhibits a reasonable shape according to the appointed contact angle of 30°, ongoing motion of the outmost liquid particles leads to unphysical covering of the solid. The drag forces of these particles lateral to the solid surface are not canceled out completely by surface tension forces. Moreover, the attractive forces induce a bad particle distribution. Without additional repulsion and only the small negative contribution of equation 5.73 for distances below l_0, effectively a dense particle double layer is formed at the outer rim.

Wetting phenomena can only be modelled, if an additional degree of density constraint with respect to the physical density is appointed (equation 5.60). A proportional factor $C^P = 0.1$ is however sufficient. Without this constraint, the simulation crashed shortly after the picture in the right frame of Figure 6.20. Clustering of particles both in the fluid bulk and at the interface to the solid phase occurs, if only the divergence of the velocity field is used as incompressibility constraint. The new approach therefore does not work without further stabilisation.

As a last result, Figure 6.21 shows final droplet shapes, when the original force of Tartakovsky and Meakin (2005) is used (with the same interaction co-efficient of 30). The contact angle was set according to the potential based relation 5.66 of Kondo et al. (2007). The droplets are deformed and do not exhibit a round shape. Likewise, the contact angle is only to some degree achieved. Additionally, particles penetrate into the solid surface (the left and middle frame, in which there is a stronger attraction to the solid phase). Some of these shortcomings can be avoided - penetration for instance by an additional repulsive term -, but the wetting behaviour remains unrealistic. The new concept of anisotropic, surface-lateral forces is therefore more suitable to calculate surface tension and wetting behaviour on the micro scale despite its undoubted disadvantages.

Numerical results of the surface tension and wetting behaviour are sobering. The standard approach of isotropic forces is insufficient in ISPH. The reported results of Tartakovsky and Meakin (2005) appear better, which may be due to the van-der-Waals equation of state in their WCSPH approach. Yet, the applicability of different contact angles has not been shown and the scaling behaviour remains uncertain. Moreover, they had to obtain the surface tension related pressure inside the drops by a cumbersome routine, as the background pressure is always very high when using isotropic forces. The surface-lateral approach alleviates some of these shortcomings. However, there seems to be a "sweet spot" concerning the combination of particle size and interaction strength, in which it performs better than for other resolutions. It does not appear feasable to adapt the force and its constituion of respulsion and attraction when changing the resolution. A revision of the concept is therefore necessary. Generally, it is a difficult task to make the net force per unit area independent from the resolution and at the same time preserve the behaviour in the vicinity of discontinuities which become more local when the resolution is refined.

Figure 6.21: Final geometry of sessile drops at contact angles of $30, 60$ and $90°$ calculated with the literature approach of Tartakovsky and Meakin in conjunction with ISPH.

7. SIMULATION OF STRUCTURE EVOLUTION DURING DRYING

The SPH model presented in the previous chapters will be applied to single droplet drying of a suspension in the following. A simulation of the first drying period will be presented shortly. The second drying period will be simulated using various implementations of crust formation. Thereafter, the outcome of structure development will be investigated depending on adjustable parameters. The applicability of the present model will be discussed further concerning variations of gas temperature and different spatial resolutions in the discretisation. Finally, drying of a microporous structure will serve as an example for the diffusion driven drying approach. All simulations were carried out in two dimensions.

7.1 Simulation of the First Drying Period

After a short time of heat-up, a droplet reaches a quasi-equilibrium of heat transfer and cooling by evaporation. The drop temperature stays constant and heat transfer is directly converted into evaporation. During this constant rate period, the squared droplet diameter decreases linearly as (comp. sections 2.5.2 and 3.2.1)

$$\frac{dR^2}{dt} = 2R\frac{dR^2}{dt} = -R\frac{D^G}{2R}Sh\frac{p_v^{surf} - p_v^\infty}{\Re\bar{T}}\frac{MW_L}{\rho_L^0} \approx const. \qquad (7.1)$$

The radius cancels out and the other values are - in a quasi-steady-state - virtually constant. The first drying period was simulated for a droplet of 75 μm initial radius and completely unsaturated gas of 100 °C at Nusselt and Sherwood values of 4. The droplet was discretised with 6925 particles of $l_0 = 1.6\,\mu$m in hexagonal alignment. Density and other specific values were those of water, for both the liquid and the solid phase. The interaction parameter s_{ij}^0 was set to 20, less for reasons of a physical derivation, but due to the increase in computational load, when high values are chosen and the time step size needs to be decreased as a consequence of large forces/accelerations. This remains as a challenge as the course of drying runs over a much longer period of time than the typical scale of effects like surface tension and wetting.

Figure 7.1 shows the initial geometry and the simulation at 0.4 and 1 s. The graphs below show the droplet radius and its squared value over time. Drying in the SPH model takes place according to the d^2 law throughout the first 0.25 s. After this, the droplet surface is more and more covered by solids and the liquid is less exposed to the gas. The drying rate is diminished with the receding of the liquid surface. After a complete layer of solid particles has accumulated at the droplet surface, evaporation virtually stops, as the LDF mass transfer only applies to liquid surface particles.

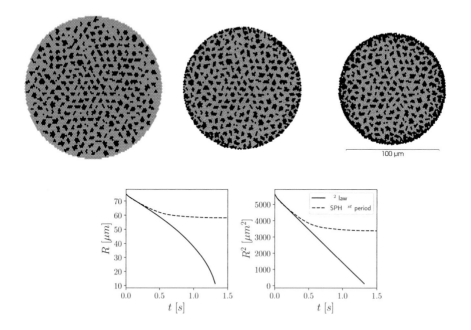

Figure 7.1: Simulation of the first drying period at 0, 0.4 and 1.5 s (black: solid, light grey: liquid) and evolution of droplet radius and squared radius over time.

7.2 Simulation of Crust Formation and the Second Drying Period

7.2.1 Simulation of the Second Drying Period without Crust Formation

Accumulation of solids at the droplet's outer rim and reduced evaporation rates result in an increase of temperature up to the point that the drop is boiling. Additional uptake of heat in the liquid phase will be recalculated to evaporation so that all liquid particles remain at 100 °C (see section 5.7.3). This results in further droplet shrinkage until full evaporation is achieved. The course of drying is depicted in Figures 7.2 and 7.3, which show the evolution of the droplet and its constitution along with the temperature distribution. The gas temperature was set to 300 °C in this simulation, Nusselt and Sherwood numbers to 4. Initially, the droplet was slightly below the wet bulb temperature so that it is heated up at first. After partly accumulation of solid particles at the surface (second frame), the free water surface has decreased so that cooling by evaporation is less pro-

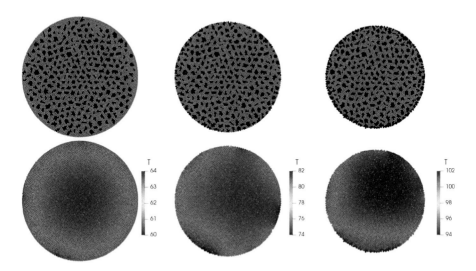

Figure 7.2: Simulation of the second drying period without crust formation: suspension (black: primary particles, grey: water) and temperature profile in the drop at times 5, 75 and 125 ms.

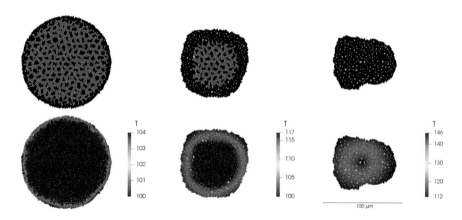

Figure 7.3: Simulation of the second drying period without crust formation at times 175, 250 and 286.5 ms.

nounced and the overall temperature is higher. Larger local solid coverage of the surface is partly visible by higher temperatures. In the third frame, the droplet has partly exceeded the boiling temperature in regions where its surface is fully covered by solids. This is the case for the overall surface in Figure 7.3, which shows the progress of boiling up to a completely dry product. Figure 7.4 shows the initial and final structure of the drop. Colours indicate the index of the rigid bodies that represent the solid phase, which can be used in order to distinguish between the different primary particles in the final geometry. It is well visible that the final structure was obtained by shrinkage without internal mixing.

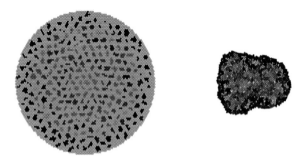

Figure 7.4: Initial and last constitution of the droplet without crust formation, colours indicate the "body index" of the suspended primary particles.

7.2.2 Crust Formation by Caught on First Touch

The previous simulation showed that solids are accumulating at the droplet's surface, but that without further stabilisation the structure is just densified. The most simple approach of implementing crust formation is to merge primary particles as soon as their minimum distance undercuts a certain value. In Figure 7.5, the particle spacing l_0 was used. Colour coding indicates again the index of the rigid bodies. The first frame shows the situation, when already some particles have merged. Relatively large clusters of irregular structure consist of several merged bodies. Throughout the first row, merging leads to a rigid crust covering the whole droplet. Subsequent drying results in the formation of a bubble inside the solidified outer structure. After complete drying, a granule is obtained, which consists of a microporous crust. Single solid particles remain unmerged as their distance is slightly above the threshold value. In a productive simulation suspended solids of one single particle size should be avoided, not only for this kind of behaviour, but also in order to provide a rather continuous approx-

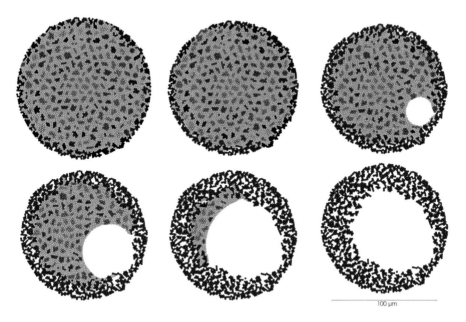

Figure 7.5: Crust formation by the caught on first touch algorithm: instants of time: $125, 150, 175, 200$ and 311.5 ms. The colour coding shows the index of the suspended solids.

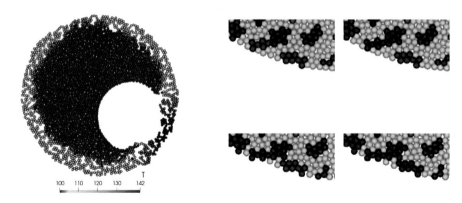

Figure 7.6: Crust formation by caugt on first touch: temperature distribution (left) and artifacts of the algorithm with primary particle merging in presence of surrounding liquid (right).

imation. The temperature distribution at the time of bubble growth is drawn in Figure 7.6 on the left. The water within the crust has reached the boiling point, whereas the solid outer structure is further heated up. The approach represents production of a hollow granule by spray drying in a basic manner. Despite exhibiting the process of structure formation in principle, the algorithm has its drawbacks and leads to early primary particle merging. Artifacts of the algorithm are depicted in the right graphs of Figure 7.6. Solids are merged, which are fairly covered by liquid particles, especially in the vicinity of the merging point. The caught on first touch algorithm therefore involves unphysical agglomeration. This is also the reason, why the final structure is very porous. Compaction of the solid layer is not possible. As soon as primary particles encounter each other, they will merge if their distance undercuts the threshold value. Adaption of the threshold to lower distances can partially act as a remedy, but local variations in the SPH particle distribution may still involve irregular primary particle merging. Moreover, adaption of the threshold distance of SPH particles in order to control the solid hull structure is not motivated by physics. It is therefore advisable to control crust formation by properties with relevance to drying.

7.2.3 Crust Formation Determined by the Water Content

Primary particles might be "caught on first touch" even in the liquid bulk. The algorithm does not account for drying at all, but just merges solids based on a coarse distance criterion. The local degree of drying can be evaluated according to the local liquid content of a particle (see section 5.6.2). The simulations in Figure 7.7 employed a limit value of zero, i. e. two solids do only merge if the neighbourhood of their closest particle pair solely consists of other solid particles. Compared to the previous result, primary particle merging is slowed down and only sets in, when drying has advanced. The fourth frame shows the last output before full merging of the crust with bubble formation starting little later. This approach is much more physical than the naive caught on first touch algorithm and provides at the same time a adjustable parameter which can be used to simulate structure variations.

The point of bubble formation is different to the previous example. Generally, the inner bubble occurs reproducibly in repeated simulations, but randomly concerning parameter or geometry variations.

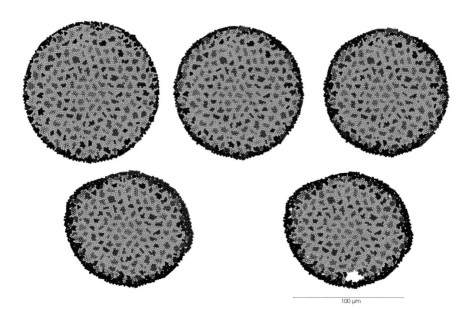

Figure 7.7: Crust formation depending on the water content (150, 175, 187.5, 199.5 and 202.5 ms).

7.2.4   Effect of the Density Correction

Figure 7.8 shows the density evolution within the droplet at the time of bubble formation. The upper row refers to a simulation in which the additional density constraint in equation 5.60 was considered with a proportional factor $C^P = 0.1$. It becomes visible that due to drying below the crust and due to interparticle forces filling gaps, the fluid is "elongated" like a rubber. When finally at some point the liquid bulk detaches from the solid crust - starting in the second frame -, this process happens very suddenly. The liquid returns to its physical density within 1 ms. Hence, the bubble expands very fast, just as "the rubber snaps back". Further bubble growth is much slower, according to the drying rate. This unphysical behaviour is avoided, if C^P is set to one. The lower row shows that bubble growth takes place in a physically correct way. Directly after bubble initiation, water fills unoccupied pore space, when the bubble allows the liquid to relax. Bubble growth is hence rapid for a very short, initial amount of time and continuing more moderate thereupon. Without or with only a small density correcting constraint in the PPE, unphysical liquid "stretching" retards bubble formation. If finally a bubble has formed, it becomes large very quickly due to the previous delay in its formation. For this reason, the density correction was applied in other drying simulations involving a crust with a proportional factor of one.

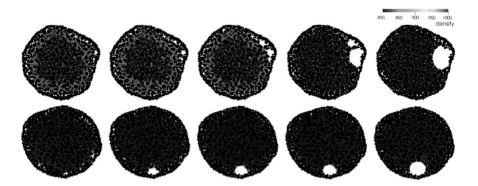

Figure 7.8: Effect of the density correction, upper row: proportional physical density constraint of 0.1, lower row: full additional contribution of the physical density constraint ($C^P = 1$), times for each example: $0, 0.5, 1, 1.5$ and 3.5 ms from the instant of bubble formation.

7.3 Influence of Adjustable Parameters on the Simulated Structure

With the maximum water content and the minimum solid content for merging, $\varphi_{lim}^{L,merge}$ and $\varphi_{lim}^{S,merge}$, and the parameter for solid-solid interaction compared to liquid-liquid attraction, f_{SS}, there are several options to adjust the simulation. In the following the term "granule" will be used for the final structure in order to distinguish it from SPH particles and primary particles. The setup was resolved finer than in the previous simulations with $l_0 = 0.8\,\mu m$ and 27619 particles. Other parameters were chosen as before.

Figure 7.9 shows variations of $\varphi_{lim}^{L,merge}$ and $\varphi_{lim}^{S,merge}$ at constant solid↔solid interaction strength. The limiting water content has an influence on the granule size. The smaller the limit is, the smaller the final granule size. Moreover, the final structure appears more dense for a small limit value. This can be well understood from the model. A higher limit value will allow crust formation in presence of remaining water, hence at an earlier instant of time. Also the tendency to cluster two primary particles is higher. At a value of 0.15, one can even observe some merging below the surface. This is similar to the caught at first touch algorithm and 0.15 is not too far away from this kind of behaviour.

Increase of $\varphi_{lim}^{S,merge}$ leads to a more dense structure. Also the granule size appears smaller and less round / more deformed. This can also be explained as a natural model behaviour. As it takes a higher solid content to combine, primary particle merging is retarded. Suspended solids need to advance further in order to cluster. Moreover, some larger clusters may first need to change their orientation before they exhibit a point of contact at which the solvent content is large enough. There is no distinction between $\varphi_{lim}^{S,merge} = 0$ and $\varphi_{lim}^{S,merge} = 0.1$ as the regarded solid SPH particle itself contributes to the summation term so that in additional neighbourhood of another solid this value is always exceeded.

It is desirable to find an equivalent from real structure generation to these values. Eckhard et al. (2014) observed in spray drying experiments an increase in microporosity at an elevated binder content. Furthermore, they proposed the mechanism that droplet shrinkage will stop earlier, when the binder content is increased so that larger granule sizes are obtained. In that respect, the limit liquid content has a similar effect - a higher value allows earlier merging and therefore acts the same way as an increased binder content. The increase in microporosity is visible from the graphs likewise. The solid threshold value may

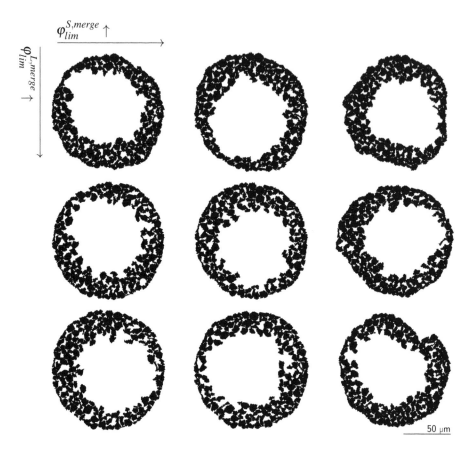

Figure 7.9: Different granule structures by model parameters: variation of $\varphi_{lim}^{L,merge}$: 0.000, 0.075 and 0.150 (rows) and $\varphi_{lim}^{S,merge}$: 0.10, 0.25 0.40 (columns); $f_{SS} = 0.5$.

be regarded similarly, but in the other direction. A lower solid limit promotes merging and thus the formation of stable clusters, whereas a larger number retards this process. In a preliminary way, an increase in binder content could therefore be reflected in the model as starting with a large limit value for the solid content and zero for the water fraction, decreasing first the solid threshold and, when this value is at its lower limit, afterwards increasing the limit for the liquid threshold.

Figure 7.10 contains a variation of liquid fraction limit and the solid-solid

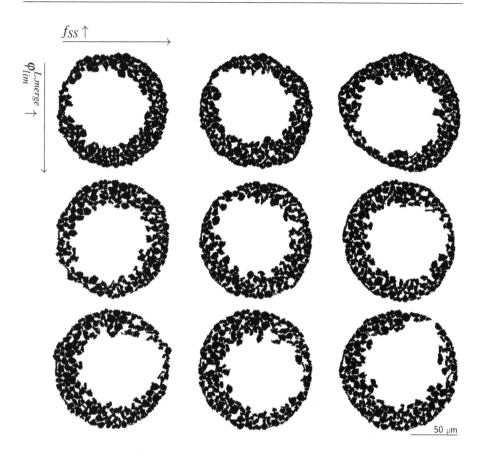

Figure 7.10: Different granule structures by model parameters: variation of $\varphi_{lim}^{L,merge}$: 0.000, 0.075 and 0.150 (rows) and f_{SS}: 0.0, 0.5 and 1.0 (columns); $\varphi_{lim}^{S,merge} = 0.10$.

attraction at constant $\varphi_{lim}^{S,merge}$. Besides the previously discussed relation concerning the limit water content, structural differences are not visible. The interaction between solids within the present surface-lateral force approach is too low to play a role.

The proposed sequence of $\varphi_{lim}^{L,merge}/\varphi_{lim}^{S,merge}$ combinations was applied to a different initial geometry in the simulation in Figure 7.11. The succession of parameters works fairly well in changing the structure from a porous to a dense shell. $\varphi_{lim}^{S,merge} = 0.60$ leads to a dense structure, but folded to some degree. For obvious reasons, particle merging will stop at some point, when the limit

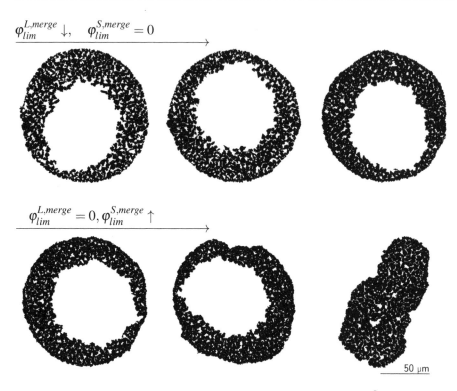

Figure 7.11: Different granule structures with decreasing $\varphi_{lim}^{L,merge}$ and increasing $\varphi_{lim}^{S,merge}$, upper row: $\varphi_{lim}^{L,merge} = 0.150, 0.075, 0, \varphi_{lim}^{S,merge} = 0$, below: $\varphi_{lim}^{L,merge} = 0, \varphi_{lim}^{S,merge} = 0.2, 0.4, 0.6$.

is too high. Yet, a value of 0.6 still allows some primary particles to merge. A completely dense structure is obtained for yet higher values. Indeed, the simulation in section 7.2.1 of the second drying period without crust formation was controlled by an impossible value of $\varphi_{lim}^{S,merge}$. This also supports the interpretation of both limit values as an analogon to the binder content, with absence of a binder denoted by high values of $\varphi_{lim}^{S,merge}$. In a further advanced model, the binder content could be implemented as an additional particle property, which is subject to diffusion. By the local binder content, the stickiness of two primary particles could be derived.

The final example in Figure 7.12 compares a suspension of small primary particles (upper row, about 2 μm diameter) with structures obtained by large primary particles (approx. 6 μm, below). Eckhard et al. (2014) observed higher microporosities for a larger primary particle size. According to their analysis, this is caused by the flow behaviour and that large particles hinder each other during drying so that the shell is rather loosely packed. This also appears within the graphs, which show a reduced microporosity in case of the small particles.

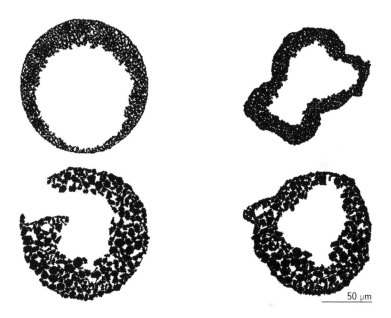

Figure 7.12: Variation of primary particle size: small (upper row), large (below), $\varphi_{lim}^{L,merge}/\varphi_{lim}^{S,merge}$ combinations of 0.1/0.0 (left) and 0.0/0.4 (right).

7.4 Effect of the Temperature

Harsh drying conditions are expected to lead to an earlier crust formation. Capillary forces are less able to pump liquid to the surface and to drag solid particles into the droplet. A stable crust should be present at an earlier stage of drying, at a larger droplet radius. In Figure 7.13, the drying gas temperature was varied between 75 and 300 °C for different $\varphi_{lim}^{L,merge}/\varphi_{lim}^{S,merge}$ settings. The previously stated dependency of the final structure on the merging threshold parameters is visible, but the granule size remains practically unchanged if the temperature is varied. Approximation of the final diameter by the granule's minimum and maximum extent in x and y direction shows virtually identical values. Even at very mild drying conditions with $T^G = 75$ °C, the droplet diameter appears similar to those at elevated temperatures. Besides the otherwise promising results, the outcome of this simulation shows that the major work package for further model development lies in a proper representation of the surface tension and wetting behaviour on the detailed scale. A small structural effect is visible in the last row, in which the minimum solid content for merging was $\varphi_{lim}^{S,merge} = 0.4$. This regime, which is prone to crumpling, provides a rather round shape for the highest temperature in comparison to the lower ones. A second simulation exhibit the similar tendency that higher temperatures may lead to a rather round shape in the simulation, Figure 7.14. In the model, crumpling takes place at the same time as crust development and is thus partially prevented by harsh drying conditions inducing a slightly earlier crust formation.

As the effect of the drying gas temperature is practically invisible, a numerical experiment was undertaken, in which drying was strongly scaled by a modified d^2 law only for the first drying stage. This was implemented by setting a predetermined drying rate, referring to receding of a flat surface (in units of m/s). Multiplication with the density yields a mass flux per unit area. This value is scaled by $\frac{R^0}{R}$ so that the d^2 law is fulfilled (fluxes in an LDF approach also scale with $\frac{1}{R}$ due to the dependency of heat and mass transfer coefficients on the droplet radius as characteristic length). Distribution amongst surficial particles is again performed by the CSF approach (multiplication with $\left|\frac{\nabla c}{|c|}\right|$, section 5.7.2). The different results are compared by the total mass $m^* = \sum_i m^{evap,i}$ of liquid within the droplet in Figure 7.15. As the time to full evaporation is very different, the instants of time of different curves have been scaled so that the x-axis is normalised. After initially strong drying, surface coverage by pri-

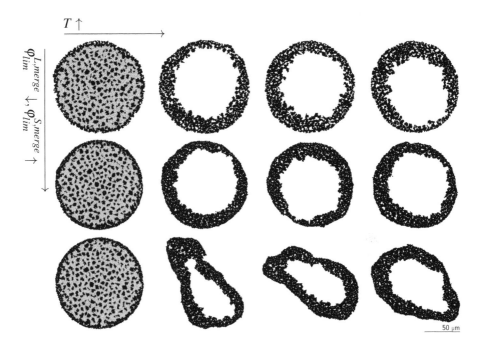

Figure 7.13: Effect of the drying gas temperature on the morphology in the model: Within each row the temperature was varied in steps of $75, 150, 225, 300\,°C$ (from left to right), first to third row: $\varphi_{lim}^{L,merge} / \varphi_{lim}^{S,merge} = 0.1/0.0, 0.0/0.0, 0.0/0.4$.

Figure 7.14: Effect of the drying gas temperature for a different geometry: In comparison to Figure 7.13 only the temperature is varied at constant $\varphi_{lim}^{L,merge} = 0.0$, $\varphi_{lim}^{S,merge} = 0.4$.

257

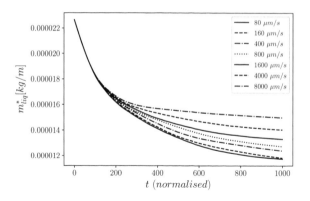

Figure 7.15: Strong variation of drying conditions by a modified d^2 law at different receding rates of the surface.

mary particles hinders further evaporation, because only the first drying period is modelled. With full surface coverage, the curves approach a horizontal tangent. This happens the earlier, the higher the drying rate is. The principle physical behaviour is hence represented by the modell, yet only for very extreme variations. The lowest and the highest curve differ in two orders of magnitude of the drying rate, which corresponds with a temperature difference between droplet and gas being 100 times higher (referring to the quasi-equilibrium of heat and mass transfer, in which the heat flux from the gas directly determines the drying rate).

The reason that this effect is so little covered lies in the current weak spot of the model, the implementation of surface tension and wetting. This result had been obtained in an intermediate model stage using isotropic forces, but the outcome would not be improved with anisotropic, surface-lateral forces as the temperature variation showed previously. The model not only suffers from a lack of dynamics and scaling in wetting, which directly affects capillary pumping, it is also a matter of resolution and the restriction to two dimensions. When two suspended solids approach each other at the drop's surface, the gap between those primary particles becomes very narrow. At some point only a single layer of liquid particles remains in between. This single particle is indeed pulled towards the surface, but the force is not high enough to drag a chain of liquid particles through the gap. This is also due to the reason that the wetting forces along both curved primary particles forming the gap point to partly opposite directions. The liquid particle needs to "decide", on which solid's surface it should move along by a slight difference in attraction. Hence, the net capillary force becomes small so that the particle is effectively stuck. Concerning such particle alignments, it must be clearly stated that the setup runs out of the typical applicability of SPH, which still is a continuum based method - the term "smooth" is already indicated in its name. This issue will for some part be overcome by a finer resolution, but remains in principle as long as surface tension and wetting effects are not incorporated in a better way and as long as simulations are carried out in 2D.

The problem of two-dimensionality could be treated by introducing an additional diffusivity through primary particles. By this, liquid might be transported to the remaining particles in the gap so that evaporation of these particles is less effective and they remain longer at the surface. This would retard crust formation. As a second variant, the diffusing liquid would not be attributed to other particles but evaporate at the solids surface. In doing so, the temperature in the vicinity of remaining liquid particles at the surface would become smaller and thus diminish their drying rate. Mass loss would be transferred to the inner part of the droplet so that shrinkage and on the other hand shell growth from the inside would be more realistic.

7.5 Variation of the Resolution

The validation examples of wetting and surface tension showed a significant effect of the SPH particle size on the physical effects. From Figure 7.16 this does not appear to be such a distinct effect, when structure evolution is simulated. One reason for this is that the influence of the wetting behaviour is generally underestimated in the model as the previous example of temperature variation showed. The final structures exhibit a similar shape and porosity and are mostly equal in size. This is also indicated by the time of complete drying, which is very close together for most structures and mostly scatters within less than two percent (Table 7.1). As the amount of total heat transfer per time depends on the surface, hence the radius, this is an indicator for a similar evolution of surface over time. The three coarsest and the finest resolution exhibit a slightly longer drying time, indicating random scatter in a fairly low range and not a systematic behaviour. Whilst this kind of study has not been performed on a large number of geometries, the outcome of this variation is yet promising.

Table 7.1: Time until full evaporation

l_0 [μm]	particle nr.	t_{dry} [s]	l_0 [μm]	particle nr.	t_{dry} [s]
0.9	21823	0.2840	1.5	7861	0.2750
1.0	17695	0.2780	1.6	6925	0.2745
1.1	14599	0.2755	1.8	5461	0.2790
1.2	12283	0.2765	2.0	4429	0.2860
1.3	10435	0.2830	2.2	3643	0.2825
1.4	9019	0.2750	2.4	3067	0.2900

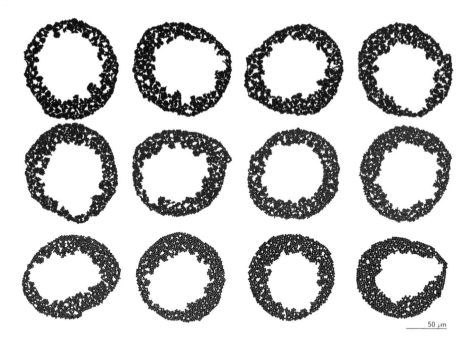

50 µm

Figure 7.16: Variation of the SPH particle resolution: particles' length scale l_0 varied from 0.9 to 1.8 µm in steps of 0.1 µm, additionally 2.0, 2.2 and 2.4 µm; $\varphi_{lim}^{L,merge} = 0.0$, $\varphi_{lim}^{S,merge} = 0.0$.

7.6 Drying of a Microporous Structure

The final examples concern drying driven by diffusion in the gas. The principle has been laid out in section 5.7.5. The domain is flat and periodic in x-direction. It is bounded below by a solid layer. Solid objects of different size are irregularly distributed within the domain and spatially fixed. These solids are covered by a water layer, which is exposed to air at its upper side. The upper air boundary is set to a constant vapour pressure of zero. Discretisation of solids and liquid is performed by SPH, the gas is implemented as an underlying grid.

Figure 7.17 shows the initial model state. The domain size was 160 times 80 μm. The particle spacing and the mesh size of the gas were set to 1.33 μm. The temperature was 60 °C and the diffusion coefficient in the gas chosen to $3 \times 10^{-5}\,\mathrm{m^2/s}$. Values of the liquid phase were set to those of water. Gravity was neglected, as surface tension and wetting dominate on the micro scale.

The course of a simulation is shown in Figure 7.18 for 0° contact angle. After initial evaporation of the upper liquid layers, wetting effects begin to play a role. The gap over the periodic boundary is depleted first, as the other distances between solids are smaller. For a longer period of time, the liquid remains in upper, rather narrow gaps. The evaporation rate is increased, as capillary pumping constantly drags towards the upper solids, where the gradient for diffusive vaporisation is higher. With receding water, evaporation is slowed down.

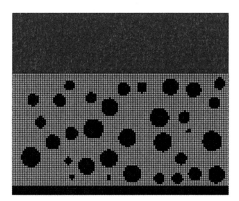

Figure 7.17: Initial setup for diffusive drying of a microstructure: fixed solid particles (grey) and water (blue) discretised by SPH, the gas (blue layer) by an underlying grid.

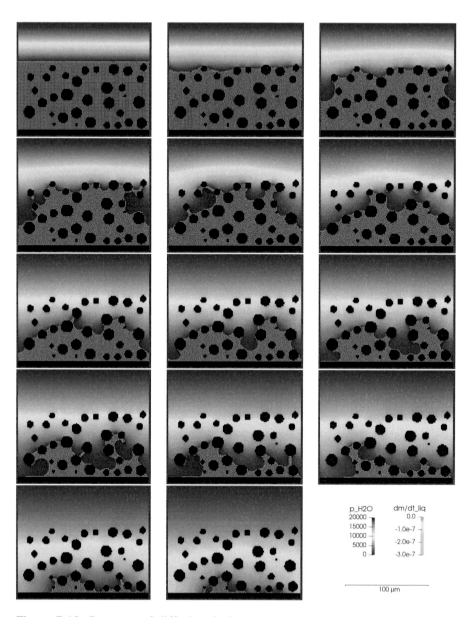

Figure 7.18: Progress of diffusive drying over time in steps of 100 ms, except for the first frame at 2 ms; colour coding of liquid particles: evaporation rate, gas: vapour pressure.

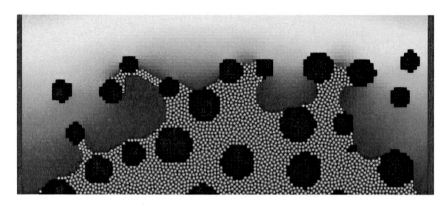

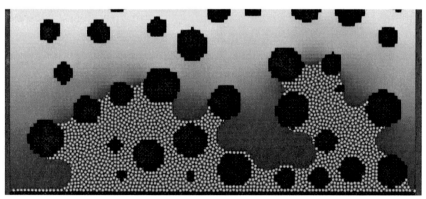

Figure 7.19: Detailed pictures of diffusive evaporation at 400 and 800 ms, colour coding of liquid particles according to the evaporation rate, gas corresponding to the partial pressure of water.

Figure 7.19 shows the process in more detail for the frames of 200 and 400 ms. The colour coding of the liquid depicts the evaporation rate. It is well visible that evaporation is mainly happening at the upper liquid surface. The vapour pressure in the gas phase is higher in the lower regions and therefore the driving force for evaporation diminished, especially, when the gas is partially trapped in pores. As the diffusive distance is smallest at the upper water surface, water is constantly removed there. This evokes a continuos flux of liquid from below and the side to the upper region due to capillary pumping. The model is quite capable of reflecting this behaviour, which in particular becomes visible in an animated view showing continuous particle motion towards the surface. Some minor issues remain. For one thing, there is a tendency to form long chains of three particles width. If a particle within this chain is removed by evaporation, pairwise forces and incompressibility exert a strong drag to close the gap within this chain immediately. The thickness of such a filament is typically three particles, reflecting the neighbourhood radius. This appears as a typical behaviour of pairwise forces (the same mechanism can be observed for isotropic forces). As these forces are zero for the particle spacing of l_0 and attractive for larger distances (compare formula 5.73), removal of a particle will suddenly leave open space so that long range attraction between both surficial neighbours besides the newly formed gap is no longer outruled by a blocking particle in between. A second peculiarity is visible in the lower picture. Along the solid surface there is a chain of particles which covers this surface. Break-up of the chain due to receding liquid is strongly retarded in the simulation. This is in principle the analogon to the long liquid filament just with the solid phase involved. It exhibits the same tendency to close a gap - or in this case rather no to let the chain be opened. This observation may be helpful for the future definition of pairwise forces and their dependency on the distance, as it displays the effect of single particle pairs' contributions.

The water position for different contact angles after 800 ms is drawn in Figure 7.20. It is not surprising that 0 and 30° contact angle come out similarly, as the discrete particle structure is not able to resolve such low angles. The water distribution at 90° wetting angle is however surprising. The picture shows again a long and stable water filament in the upper middle. It is such an irregularity which exerts a strong drag on the rest of the liquid. With the regime being only partially wetting otherwise, such a filament continuously drags water out of the rest of the pore structure. It is also visible that this really is an artefact as the

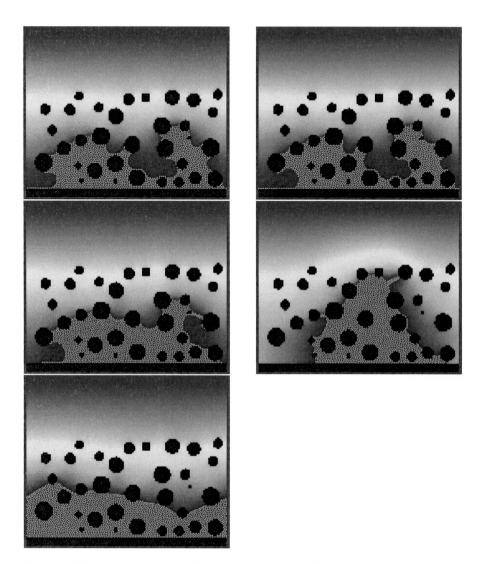

Figure 7.20: Comparison of diffusive drying at 800 ms, contact angle in the graphs (from upper left): $0, 30, 60, 90, 120°$.

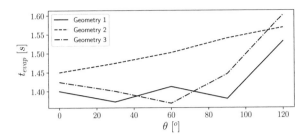

Figure 7.21: Time to full evaporation vs. contact angle for three different geometries.

curvature is unrealistically small at the point, at which the filament goes into the liquid bulk. The liquid with 120° contact angle is rather pushed downwards so that the diffusive length is effectively maximised. Comparing the times until full evaporation for three different geometries (7.21), the model shows in average the behaviour that with smaller contact angles water is more likely to be dragged towards the surface and evaporates faster. The dynamic at small contact angles is however not satisfactory as already discussed.

The approach of a coupled SPH-grid model is an efficient way to integrate a third phase, especially, when equations on the grid can be solved implicitly at large time steps. To the author's knowledge this is its first application to vapour diffusion and evaporation. Besides the already discussed deficiencies of the surface tension approach, the model represents the underlying physics very well.

A further potential of SPH-grid coupling lies in the calculation of a smooth colour function over both grid and particles for a CSF implementation of surface tension forces in a free surface SPH solver. Even an extension to wetting phenomena does not seem out of reach. This could be done by treating the grid as SPH particles like in the diffusive coupling in the present model. Alternatively, the SPH coverage of the grid could be used in order to determine the boundary according to a volume of fluid (VOF, Hirt and Nichols 1981) reconstruction for which the CSF approach is well established. The final forces might then be backtransformed to SPH.

7.7 Comments on numerical efficiency

SPH is a numerically demanding method. Simulations of finely resolved structures with initially more than 27000 particles took about two days in a serial code on Core i7 machines. Parallelisation of the method is possible and has been performed many times before. Calculation of rigid bodies over different compute nodes is a bit tricky, but not an unresolvable task. Yet, an extension to 3D will increase the computational load tremendously as it both involves a much larger number of particles and a strongly englarged discretisation stamp. Calculation of the complete drop in 3D is therefore out of the reach. Structure simulations in three dimensions should rather be applied to representative parts of the overal domain.

Moreover, the coupling of rather slow (drying) and fast (surface tension) physical effects is challenging. Further development of the method should therefore concern ways of an increased computational efficiency.

8. CONCLUSION

Within this work two new single droplet drying models have been developed. In the first part an enhanced model of solution drying has been derived for the calculation of polymerisation in drops. Chemical reactions are either considered by the method of moments or the simpler quasi-steady-state-assumption. Special regard needs to be paid on conservation of mass in the reaction-diffusion system and on the proper application of moments' / polymer density diffusion. The literature approach converges to a uniform polymer of constant chain length. Based on Maxwell-Stefan equations a new approach was derived, which retains spatial gradients in polymer properties. The model reveals new insight into spray polymerisation processes. The previously proposed concurrency of drying and chemical reaction was not confirmed. Drying reduces the temperature within the drop that much that chemical reactions scarcely take place. Polymerisation therefore happens for the largest part in bulk, nearly independent from the solvent content in the feed. Moreover, evaporation of monomer strongly decreases the yield, if not countered by additional measures. If the monomer saturation in the drying gas is elevated in order to circumvent this behaviour, reactive absorption of monomer takes place. Differently to drying, the process does not require a high amount of energy as long as a certain level of polymer yield is achieved. As the process happens for the largest part under (near) bulk conditions, modelling becomes partly uncertain, because many literature kinetics are measured within solvent. The simulation results are therefore to some degree speculative and only qualitative statements about the process could be made within this work.

A deeper analysis of process variants by means of numerical DoEs showed that spray polymerisation in presence of a solvent is scarcely applicable, as the solvent content within the drop is strongly lowered by evaporation. As the reactions are performed in bulk anyway, it appears reasonable to leave out the solvent as long as precipitation is not a concern. Two process variants for bulk polymerisation appear feasable. If the drying gas is just recirculated, the monomer content within the gas will adapt, until the balance between monomer evaporation and uptake by reactive absorption in droplets is in an equilibrium. By this, a yield of 100 % can be achieved theoretically. This, however, involves large monomer saturations in the drying gas with additional challenges like the possibility of thermal initiation or condensation at cold spots. Furthermore, reactive absorption needs to be stopped before the droplets leave the dryer, which may involve additional steps. The second alternative is to set the monomer saturation in the gas to a lower value, thus avoiding unwanted side effects at the cost of a significantly lower yield and an additional step of monomer recovery from the gas. Slight pre-polymerisation before atomisation induces a skin of low permeability at the droplet's surface, which reduces monomer evaporation and thus increases the yield to some extent. This involves further challenges like atomisation of a polymer rich solution and prevention of clogging. The findings within this work are of general nature and reveal most promising paths for further experiments and research on spray polymerisation. Still, the conclusions need to be confirmed experimentally.

In the second part, a novel approach of single droplet modelling of suspension drying has been derived, which aims at a direct calculation of structure evolution during drying by incorporation of the underlying physics on a detailed scale. This was realised by the meshfree SPH method, which has not been applied to droplet drying applications until now. Therefore, basic concepts like the incorporation of linear driving force-based heat and mass transfer into the method have been derived. The final model takes the hydrodynamics of an incompressible fluid consisting of a liquid phase and solid primary particles, surface tension and wetting effects and heat and mass transfer into account. The second drying stage has been implemented as droplet boiling. The evolution of a crust at the drop's surface has been considered by merging of primary particles in close contact to larger crust segments. Whereas simple merging based on a threshold distance between particles produces very porous structures, the model behaviour can be adjusted with local conditions at the point of contact between primary particles. Connecting crust solidification to a maximum local water or a minimum solid content introduces drying related parameters, which adjust the final structure obtained by simulations. The dependency of predicted product morphologies on these parameters can be interpreted in a physical manner by the binder content and reflects experimental findings. An additional example showed the feasibility of the approach to model diffusive drying within a porous structure. Diffusion in this case has been modelled by addition of the gas phase as an underlying grid. The combination of meshfree and meshbased methods is efficient and enables the SPH method to model further problems. The weak point in the current model proved to be the approach of surface tension and wetting, which has been implemented by an atomistic view of attractive and repulsive forces. While the literature approach involves net forces, which do not add up to zero on flat surfaces, the newly developed method of surface-lateral forces corrects this behaviour and leads to significantly improved results. Still, scaling the forces over different numerical resolutions remains an issue and the particlish nature of atomistic forces in a smooth continuum method is a general drawback of this approach. This point needs to be concerned in further work on this topic. Nevertheless, the presented model provides a novel approach in single droplet drying models, which allows for a detailed simulation of structure evolution and underlines the potential to simulate the morphogenesis of particulate powders in droplets from first principles.

Appendix

A. NUMERICAL REGRESSION BY GAUSSIAN PROCESSES

A Gaussian process is a generalisation of the Gaussian probability distribution (Rasmussen and Williams 2006, p. 2), which covers features of functions. In simple words, if a value y depending on x is predicted at a certain value x_p, the corresponding value y_p can be considered as the mean of a Gaussian distribution in y direction with confidence bounds according to the model's standard deviation. Hence, with a probability according to a Gaussian distribution, y values different to y_p are possible at x_p as well. In the general view of a Gaussian process, multivariate Gaussians are applied in which the average $\vec{\mu}$ is a vector and the standard deviation of the one dimensional distribution is replaced by the covariance matrix Σ. In this covariance matrix the relation between neighbouring points $\vec{x}$ and $\vec{x}'$ is expressed as a function of their distance, the kernel function $k(\vec{x}, \vec{x}')$. It appears reasonable that with increasing distance the linkage between two points should become monotonically weaker, which is for instance the case in the popular radial basis function (RBF) kernel. Problems with cyclical phenomena may be approximated well using periodic kernels though. An extension to regression on multiple input values just involves calculating the distance between $\vec{x}$ and $\vec{x}'$ in a multidimensional parameter space.

With onedimensional training data (X_t, y_t) provided, the covariance matrix

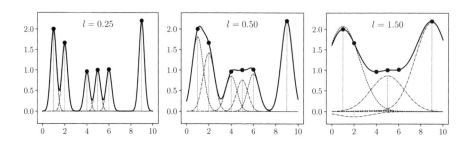

Figure A.1: Example of Gaussian process approximation via RBF kernel with varying length scale of the kernel function (no noise in the GP approximation).

K is calculated as $K_{ij} = k(X_{t,i} - X_{t,j})$. A predicted value at position x_p is then obtained by calculating the vector of the kernel function between x_p and all training points X_t (each vector row i contains $k(x_p, X_{t,i})$) and subsequent multiplication with the inverse of the covariance matrix and a vector containing all training y_t values:

$$y_p = k\left(x_p, \vec{X_t}\right) K^{-1} \vec{y_t}. \tag{A.1}$$

Figure A.1 provides simple Gaussian process approximations for a one dimensional distribution of a value with variation of the kernel's length scale. The solid line is the approximated function, whereas dashed and dashed-dotted lines refer to the contributions of the single training points. These curves are calculated with respect to the points $X_{t,i}$ using $k(x, X_{t,i})$ multiplied with the i-th element of $K^{-1}\vec{y_t}$. As the latter one solely consists of constant values - the covariance matrix K and the training data y_t, the individual curves just depend on $k(x, X_{t,i})$. Obviously, these single contributions have a Gaussian shape as well, which is a kernel property and not a common feature of a Gaussian process. The underlying RBF kernel is essentially a Gaussian itself, but other kernels would provide different shapes. With a very short kernel length scale, only a small vicinity around training points is approximated correctly whereas the predicted values tend to zero for distant points in between. However, the training data is predicted perfectly with $R^2 = 1$ and the inadequacy of this approximation would not be visible in a $y^{predicted}$ over $y^{training}$ plot. The best kernel parameter(s) are typically detected by optimisation/maximisation of the marginal likelihood of the Gaussian process approximation.

Typical data contains errors which needs to be considered by additional noise

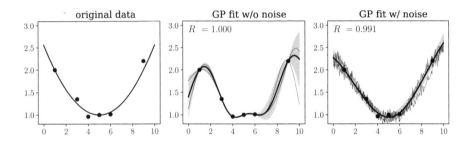

Figure A.2: Example of Gaussian process regression: original function and sampling points with errors (left), radial basis function approximation without (middle) and with noise (right).

in the kernel function. This is realised by addition of σ_n^2 to the diagonal elements of K, when σ_n is the noise level. Figure A.2 shows automated fitting of Gaussian processes to optimal kernel parameters with and without noise. The left frame contains the original function, which is in fact a second order polynomical with the sample points exhibiting some error. The fit in the middle frame without noise matches the training points perfectly with R^2 being one, but exhibits a distorted shape. In the right frame, the predicted curve does not match the training data in absolute perfection (thus $R^2 < 1$), but the shape of the curve much more resembles the original data. The gray areas around the curves denote the local one sigma interval of the approximation in y direction. As the predicted values are normally distributed, the solid line represents the most probable y values. Other courses of the curves are possible based on the same Gaussian process parameters, yet with a smaller probability. Two such sample curves are drawn as dotted lines in each graph. The one sigma interval therefore contains $\sim 68\%$ of possible predicted data for the respective Gaussian process.

Whereas the necessity of error treatment is obvious for experimental data, addition of noise in Gaussian process regression of numerical DoEs is also advisable, even if the numerical data is smooth and without scatter. As the examples show, a perfect match to training data does not necessarily mean that the approximation covers the course of a value between training points properly. The more input variables are variated, the more difficult it becomes to cover the complete parameter space. Just like in regressions using high order polynomials, artificial extrema may result from overfitting. The regressed data therefore needs to be cautiously evaluated with respect to physical meaningfulness.

It is beneficial when the values in a regression are normally distributed. If this is not the case, various transformations can be used to improve the data. Regression is then applied on the transformed data. Prediction is also performed in the transformed space with a backtransformation needed in order to obtain the predicted values. Typical transformations are the Box-Cox (Box and Cox 1964) and Yeo-Johnson (Yeo and Johnson 2000) formulae. Both follow a power law, with the Box-Cox transformation being only applicable to positive data, whereas the Yeo-Johnson approach modifies the Box-Cox formula to negative values.

The numerical DoEs in this work were prepared with the maximin construction from the diversipy package (Wessing 2015) and evaluated using the scikit-learn (Pedregosa et al. 2011) package, which provides various regression and transformation methods and automated parameter estimation. The following workflow was applied in the regression of a value y on predictors x:

- Distribute x values via maximin construction.

- Simulate droplet polymerisation in $0D$ and $1D$ for all parameter combinations with output provided each $0.1\,s$. Postprocessing provides characterical values as training data for the regression.

- Apply x data (predictor) transformation on training data according to Yeo-Johnson.

- If y values are strictly positive, apply Box-Cox, otherwise Yeo-Johnson transformation on the y training data.

- Fit via a Gaussian process with an RBF kernel by the use of the standard marginal likelihood optimisation of scikit-learn with following adjustable values: RBF kernel length scale, constant prefactor before the RBF ("constant kernel") and noise added to the RBF ("white noise kernel").

- Predict values using Yeo-Johnson transformation of desired x values and backtransformation of the predicted y values.

- If desired, calculate model bounds ($95\,\% \widehat{=} y^{predicted} \pm 1.96\sigma$) within the transformed space and backtransform the obtained bounds.

Transformation of predictor values and fit/prediction can be performed within one step using scikit-learn's Pipeline objects, wheras transformation and backtransformation of the predicted data needs to be undertaken seperately, if model bounds are to be obtained, too.

B. FINITE VOLUME (FVM) IMPLEMENTATION OF THE DROPLET POLYMERISATION MODEL

The finite volume method utilises integral formulations of conservation laws (Ferziger and Peric 2007, p. 43). Values are approximated inside discrete control volumes. This enables the method, contrarily to the point approximation of finite differences, to conserve physical properties such as mass or energy. The principle course of action can be derived by considering a conserved extensive quantity Y inside a control volume V_c with $y = \frac{Y}{V_c}$ being its intensive value. The respective property changes by sources/sinks $\sigma_{y,c}$ inside the volume c as well as by fluxes $f_{y,c}^b$ over the volume's boundaries b

$$\frac{dY_c}{dt} = V_c \sigma_{y,c} + \sum_b f_{y,c}^b A_c^b, \tag{B.1}$$

where A_c^b is the area of a boundary b. If the volume changes due to grid adaptation, an additional term needs to account for the interface motion v_c^b

$$\frac{dY_c}{dt} = V_c \sigma_{y,c} + \sum_b f_{y,c}^b A_c^b - \sum_b y_c^b v_c^b A_c^b. \tag{B.2}$$

The change of an intensive property inside volume c follows from

$$\frac{dY_c}{dt} = \frac{d(y_c V_c)}{dt} = V_c \frac{dy_c}{dt} + y_c \frac{dV_c}{dt} \tag{B.3}$$

$$\frac{dy_c}{dt} = \frac{1}{V_c} \frac{dY_c}{dt} - \frac{y_c}{V_c} \frac{dV_c}{dt}. \tag{B.4}$$

The droplet can be approximated by spherical shells around a small, centred sphere as its innermost volume. A shell c is characterised by its inner and outer radius R_c^i and R_c^o, expressed by the dimensionless coordinate ξ as

$$R_c^i = \xi_c^i R \qquad R_c^o = \xi_c^o R \tag{B.5}$$

depending on the droplet radius R. Radii of adjacent shells are connected by

$$R_c^i = R_{c-1}^o. \tag{B.6}$$

If the moving boundary problem is considered as a uniformly shrinking/expanding mesh, the change of a cell volume $\frac{dV_c}{dt}$ can be expressed as a function of the droplet radius

$$\frac{dV_c}{dt} = \left(\xi_c^{a3} - \xi_c^{i3} \right) 4\pi R^2 \frac{dR}{dt} \tag{B.7}$$

$$\frac{1}{V_c} \frac{dV_c}{dt} = \frac{3}{R} \frac{dR}{dt}. \tag{B.8}$$

The interface velocity at dimensionless position ξ is $\xi \frac{dR}{dt}$. The change of a concentration $c_{j,c}$ is therefore

$$\begin{aligned}
\frac{dc_{j,c}}{dt} = {} & \frac{3}{R} \frac{\xi_c^{i2}}{\xi_c^{o3} - \xi_c^{i3}} \left[c_j v^N + J_j^N \right]_c^i - \frac{3}{R} \frac{\xi_c^{o2}}{\xi_c^{o3} - \xi_c^{i3}} \left[c_j v^N + J_j^N \right]_c^o + r_{j,c}^F \\
& - \frac{3}{R} \frac{\xi_c^{i2}}{\xi_c^{o3} - \xi_c^{i3}} c_{j,c}^i \xi_c^i \frac{dR}{dt} + \frac{3}{R} \frac{\xi_c^{o2}}{\xi_c^{o3} - \xi_c^{i3}} c_{j,c}^o \xi_c^o \frac{dR}{dt} - c_{j,c} \frac{3}{R} \frac{dR}{dt}.
\end{aligned} \tag{B.9}$$

The first two terms account for convective and diffusive transport across the inner and outer shell boundaries. Values in the square brackets need to be calculated at the boundary positions between volume c and its adjacent cells. The third term is the source term due to chemical reactions. The second row accounts for transport due to the moving mesh (matter is transported from an inner cell to an outer one, if the mesh is shrinking) in the first two terms. $c_{j,c}^i$ and $c_{j,c}^o$ denote the concentration of species j at the inner and outer shell boundary of volume c. The third term accounts for the droplet's volume change / shrinkage (if one species evaporates and the droplet shrinks, the concentrations of other species will increase). The interface motion terms and the volume changing contribution in the second row are very similar and of opposite sign. In fact, for uniform concentrations both expressions are equal and cancel out. Still, both contributions apply to different processes, as the first part accounts for cell motion across the numerical domain and the second one for cell shrinkage/expansion and its effect on intensive properties.

The prefactors to inner and outer fluxes (bracket terms with superscript i/o) and the interface motion terms in equation B.9 are constant except for the inverse of the droplet radius. The coefficients $k_c^i = -3\frac{\xi_c^{i2}}{\xi_c^{o3}-\xi_c^{i3}}$ and $k_c^o = 3\frac{\xi_c^{o2}}{\xi_c^{o3}-\xi_c^{i3}}$ can be condensed into a sparse $n \times (n+1)$ divergence operator matrix K_{div}^{FVM} in which the rows apply to the finite volumes and the columns refer to the fluxes over the volumes' boundaries, if n volumes are used. With inter-cell fluxes and the convective terms referring to interface motion being summarised in column vectors of length $n+1$ and rates of formation and concentrations in column vectors of length n, the following vectorised notation is obtained

$$\left(\frac{dc_j}{dt}\right) = \frac{1}{R}K_{div}^{FVM}\left(c_j^{conv}v^N + J_j^N\right) + \left(r_j^F\right) - \frac{1}{R}K_{div}^{FVM}\left(c_j^{mesh}\xi\right)\frac{dR}{dt} - (c_j)\frac{1}{R}\frac{dR}{dt}.$$
(B.10)

The inter-cell concentrations concerning convection (superscript $conv$) and mesh-motion ($mesh$) need to be evaluated seperately as discussed in section 3.5.3.

Based on this matrix, the reaction induced velocity can be solved in a loop following

$$v_{i+1}^R = \left(-K_{FVM_{i,i}}^{div}v_i^R - R\left(\sum_j r_{ji}^F \frac{MW_j}{\rho_j^0}\right)\right)\frac{1}{K_{FVM_{i,i+1}}^{div}}.$$
(B.11)

The zeroth element in the velocity vector refers to the droplet centre with the boundary condition $v_0^R = 0$, the zeroth elements of the rate of formation vectors to the innermost volume. The density changing reactions are considered according to equation 3.105.

Likewise, an additional volume correction can be implemented based on the constraints that $\sum_j \varphi_j \overset{!}{=} 1$ and $\sum_j \frac{\partial \varphi_j}{\partial t} \overset{!}{=} 0$:

$$v_{i+1}^{corr} = \left(-K_{FVM_{i,i}}^{div}v_i^{corr} - R\left(\sum_j \frac{\partial \varphi_j}{\partial t} + k^{corr}\left(\sum_j \varphi_j - 1\right)\right)\right)\frac{1}{K_{FVM_{i,i+1}}^{div}}.$$
(B.12)

Beforehand, the current volume fractions φ_j are calculated from the present concentrations and their changes over time from $\frac{\partial c_j}{\partial t}$ after all differential equation terms have been evaluated. The first summation term accounts for violations of conservation in the currently calculated change of concentrations over time.

This may be due to peculiarities in diffusive higher moment transport, which is in fact of convective nature as explained in section 3.5.4. Ideally, $\sum_j \frac{\partial \varphi_j}{\partial t}$ should become zero after application of this corrective contribution, apart from discretisation errors. The second term corrects the accumulated violation of conservation in the past. The prefactor k^{corr} acts as a proportional factor which controls the ratio of correction over time. A value of 100 worked satisfactorily. After calculation of v^{corr}, an additional convective term using this velocity is added to the already evaluated $\frac{\partial c_j}{\partial t}$. It needs to be stressed that such a correction should be used as a "last measure", as it implicitly fixes hidden errors in the implementation and thus makes it difficult to find and correct those. In most cases, the proposed model equations along with the implementational considerations in section 3.5 should provide good conservation without this correction.

C. IMPLEMENTATIONAL ASPECTS OF SPH

C.1 Neighbourhood Search

An SPH discretisation of a mathematical expression contains distance dependent kernel values or derivatives for all neighbouring particles within the cut-off/neighbourhood radius of a particle of interest. Due to the arbitrary particle alignment and because of particle motion the discretisation stamp is individual for each particle and not constant. Each function evaluation of a time integration hence contains a neighbourhood search, which is one of the most time consuming parts of an SPH simulation. The trivial algorithm, in which a neighbourhood test is performed on each particle with all remaining particles, is only appropriate for very low particle numbers, as its numerical effort scales by $O\left(n^2\right)$ with the total number of particles n. If the smoothing-length and hence the cut-off radius are (nearly) constant, linked lists are a very effective way to improve the efficiency of the neighbourhood search. The case of a strongly varying smoothing length is not subject to this work, but can be treated by hierarchical tree algorithms (Hernquist and Katz 1989). An additional technique are Verlet lists, which may be combined with linked lists or a tree search as well.

C.1.1 Linked List

The algorithm of a linked list has been introduced by Hockney, Goel, and Eastwood (1973, 1974) for Molecular Dynamics (see also Schofield 1973) and can be adapted easily to SPH simulations. The computational domain is divided into cells of an edge length being equal or greater than the largest cut-off radius. Each particle is unambiguously assigned to one cell. Neighbouring particles can only be contained in the same or in adjacent cells. The number of possible neighbours of a particle thus is restricted and not affected by the total number of particles anymore. Hence, the numerical effort grows with $O(n)$ (Hockney, Goel, and Eastwood 1974; Schofield 1973).

C.1.2 Verlet List

A Verlet list (Verlet 1967) is an alternative approach in order to restrict the number of possible neighbours, which have to be tested. Contrarily to a pure linked list, which only contains the cell-particle connections, the IDs of neighbouring particles themselves are stored in a list. The Verlet list itself can be generated efficiently by a linked list algorithm (Domínguez et al. 2011). Additionally, the kernel values and its derivatives can be stored in a Verlet list, which is sensible if these values are multiply used in different loops. This is the case in corrected schemes (Domínguez et al. 2011) or for ISPH. If the radius used for calculating this Verlet list is larger than the cut-off radius, the list can be applied as a pre-selection of possible neighbours in several subsequent time-steps.

The implementation of the current model combines a linked list for the neighbourhood search with the storage of repeatedly used values in Verlet lists.

C.2 Performance Aspects, Memory Alignment

The SPH neighbourhood contains a large number of connected particles, which are not constant but may change due to particle motion. Memory cells containing particles' data are therefore accessed in a non-continuous way, which generally is a disadvantage as pre-fetching of values cannot be performed with the same efficiency as for a continuous memory access. Storage of particle data in traditional C or Fortran like arrays hence allows a faster memory access compared to data being collected in high-level objects, as the particle data alignment is advantageous. This holds especially, if objects are themselves accessed through high-level data structures like vectors. The present code was written in C++ and made use of objects and vectors, but employed traditional C arrays for storage of particle values.

A linked list of the SPH neighbourhood can be used further to optimise memory access. If particles contained in single cells of a linked list are aligned in memory as well, looping over neighbouring particles will refer to memory cells which are not necessarily continuously aligned but exhibit a minimised distance in between. A regular check of the particle alignment in memory and rearrangement of data if necessary helps to harness modern machine's caching and prefetching capabilities at least partly. Especially rearrangement of the initial, non-optimised particle distribution in memory sped up function evaluations in the numerical model noticeably in the present code.

In order to provide more data in a continuous way, neighbourhood values as the kernel W_{ij}, the interparticle distance r_{ij} and the kernel derivative $\frac{dW_{ij}}{dr}$ as well as their respective vectors and the term $\frac{m_i}{\rho_i}$ can be stored in Verlet lists. This way, these values are aligned in the succession of the neighbourhood and accessed in a continuous way when looping over neighbouring particles. The values, which are subject to the respective SPH operator, are however not distributed without gaps in their memory allignment so that usage of modern CPU's vector units (the various SSE and AVX variants) is limited.

Storage of frequently used values can additionally decrease the numerical effort, if these values are costly to evaluate. As division and square-root are amongst the most expensive mathematical operations, it is advisable to store respective values within a Verlet list as well. This concerns the inverse of the interparticle distance r_{ij}, which occurs in the denominator of the Laplace operator in Brookshaw's formulation 5.23 or when interparticle forces are given a direction by $\frac{\vec{r}}{r_{ij}}$. Similarly, storing the pre-computed $\frac{m_j}{\rho_j}$ value avoids unnecessary divisions by ρ_j.

The amount of square-root evaluations within a neighbourhood search can be minimised, if the test, whether a particle j is within the neighbourhood radius, is not performed with respect to the particle distance r_{ij}, but the previously calculated value $r_{ij}^2 = \sum_k r_{ij}^{\alpha 2}$ (α denoting the spatial dimensions).

Unnecessary computations can be avoided, if within the Brookshaw *div grad* operator 5.23 not the often stated form

$$\nabla^2 f_i == \sum_j 2 \frac{m_j}{\rho_j} (f_i - f_j) \frac{\vec{r}_{ij} \nabla_i W_{ij}}{r_{ij}^2 + \eta_{div}^2} \tag{C.1}$$

is used, which involves an unnecessary scalar product of particle distance and the kernel gradient, but the mathematically equivalent form

$$\nabla^2 f_i = \sum_j 2 \frac{m_j}{\rho_j} \frac{f_i - f_j}{r_{ij} + \eta_{div}} \frac{dW_{ij}}{dr} = \sum_j 2 \frac{m_j}{\rho_j} (f_i - f_j) F_{ij}. \tag{C.2}$$

The term $F_{ij} = \frac{1}{r_{ij}} \frac{dW_{ij}}{dr}$ can additionally be precomputed and stored in a Verlet list. Dropping of η_{div}, if possible when particles do not coincide, alternatively allows for the computation with a previously stored inverse of r_{ij} in order to skip the division operation.

Bibliography

Abrams, D. S. and J. M. Prausnitz (1975). "Statistical thermodynamics of liquid mixtures: A new expression for the excess Gibbs energy of partly or completely miscible systems". In: *AIChE Journal* 21.1, pp. 116–128. ISSN: 1547-5905.

Adami, S., X. Y. Hu, and N. A. Adams (2013). "A transport-velocity formulation for smoothed particle hydrodynamics". In: *Journal of Computational Physics* 241.0, pp. 292–307. ISSN: 0021-9991.

Adami, S., X. Y. Hu, and N. A. Adams (2010a). "A conservative SPH method for surfactant dynamics". In: *Journal of Computational Physics* 229.5, pp. 1909–1926. ISSN: 0021-9991.

Adami, S., X. Y. Hu, and N. A. Adams (2010b). "A new surface-tension formulation for multi-phase SPH using a reproducing divergence approximation". In: *Journal of Computational Physics* 229.13, pp. 5011–5021. ISSN: 0021-9991.

Aly, A. M., M. Asai, and Y. Sonda (2013). "Modelling of surface tension force for free surface flows in ISPH method". In: *International Journal of Numerical Methods for Heat & Fluid Flow* 23.3, pp. 479–498. ISSN: 0961-5539.

Antuono, M., A. Colagrossi, and S. Marrone (2012). "Numerical diffusive terms in weakly-compressible SPH schemes". In: *Computer Physics Communications* 183.12, pp. 2570–2580. ISSN: 0010-4655.

Arriola, D. J. (1989). *Modeling of Addition Polymerization Systems*. Madison.

Askes, H. and S. Ilanko (2006). *The use of negative penalty functions in linear systems of equations*.

Baehr, H. D. and K. Stephan (Aug. 12, 2010). *Wärme- und Stoffübertragung*. Springer-Verlag GmbH. ISBN: 9783642101946.

Balay, S. et al. (2019). *PETSc Web page.* https : / / www . mcs . anl . gov / petsc.

Baltsas, A., D. S. Achilias, and C. Kiparissides (1996). "A theoretical investigation of the production of branched copolymers in continuous stirred tank reactors". In: *Macromolecular Theory and Simulations* 5.3, pp. 477–497. ISSN: 1521-3919.

Bathe, K.-J. (1996). *Finite element procedures.* Englewood Cliffs, NJ: Prentice Hall. ISBN: 0133014584.

Benz, W. (1990). "Smooth Particle Hydrodynamics - a Review". In: *Numerical Modelling of Nonlinear Stellar Pulsations Problems and Prospects.* Ed. by J. R. Buchler, pp. 269–+.

Biedasek, S. K. (July 1, 2009). *Aufbau eines akustischen Levitators zur Durchführung und Online-Verfolgung von Polymerisationen in Einzeltropfen als Modellexperiment.* Wissenschaft + Technik Ve. 168 pp. ISBN: 3896852213.

Bilotta, G. et al. (2011). "Moving least-squares corrections for smoothed particle hydrodynamics". In: *Annals of Geophysics* 54.5.

Bird, R. B., W. E. Stewart, and E. N. Lightfoot (2002). *Transport phenomena.* 2nd ed. New York: J. Wiley. ISBN: 0471410772.

Bøckmann, A., O. Shipilova, and G. Skeie (2012). "Incompressible SPH for free surface flows". In: *Computers & Fluids* 67.0, pp. 138–151. ISSN: 0045-7930.

Bondi, A. A. (1968). *Physical properties of molecular crystals, liquids, and glasses.* New York, NY [u.a.]: Wiley. ISBN: 0471087661.

Bonet, J. and T. S. Lok (1999). "Variational and momentum preservation aspects of Smooth Particle Hydrodynamic formulations". In: *Computer Methods in Applied Mechanics and Engineering* 180.1-2, pp. 97–115. ISSN: 0045-7825.

Bonet, J. and S. Kulasegaram (2000). "Correction and stabilization of smooth particle hydrodynamics methods with applications in metal forming simulations". In: *International Journal for Numerical Methods in Engineering* 47.6, pp. 1189–1214. ISSN: 1097-0207.

Box, G. E. P. and D. R. Cox (Jan. 1964). "An Analysis of Transformations". In: *Journal of the Royal Statistical Society. Series B (Methodological)* 26.2, pp. 211–252. ISSN: 00359246.

Brackbill, J. U., D. B. Kothe, and C. Zemach (1992). "A continuum method for modeling surface tension". In: *Journal of Computational Physics* 100.2, pp. 335–354. ISSN: 0021-9991.

Brenn, G. (2004). "Concentration Fields in Drying Droplets". In: *Chemical Engineering & Technology* 27.12, pp. 1252–1258. ISSN: 1521-4125.

Brenn, G. et al. (2007). "Evaporation of acoustically levitated multi-component liquid droplets". In: *International Journal of Heat and Mass Transfer* 50.25–26, pp. 5073–5086. ISSN: 0017-9310.

Brookshaw, L. (1985). "A method of calculating radiative heat diffusion in particle simulations". In: *Proceedings of the Astronomical Society of Australia* 6, pp. 207–210.

Brookshaw, L. (1994). "Solving the Heat Diffusion Equation in SPH". In: *memsai* 65, p. 1033.

Budde, U. and M. Wulkow (1991). "Computation of Molecular Weight Distributions for Free Radical Polymerization Systems". In: *Chemical Engineering Science* 46.2, pp. 497–508. ISSN: 0009-2509.

Chaniotis, A. K., D. Poulikakos, and P. Koumoutsakos (2002). "Remeshed smoothed particle hydrodynamics for the simulation of viscous and heat conducting flows". In: *J. Comput. Phys.* 182.1, pp. 67–90. ISSN: 0021-9991.

Chen, J. K., J. E. Beraun, and T. C. Carney (1999). "A corrective smoothed particle method for boundary value problems in heat conduction". In: *International Journal for Numerical Methods in Engineering* 46.2, pp. 231–252. ISSN: 1097-0207.

Chorin, A. J. (1968). "Numerical Solution of the Navier-Stokes Equations". In: *Mathematics of Computation* 22.104, pp. 745–762. ISSN: 00255718.

Cleary, P. W. (1998). "Modelling confined multi-material heat and mass flows using SPH". In: *Applied Mathematical Modelling* 22.12, pp. 981–993. ISSN: 0307-904X.

Cleary, P. W. and J. J. Monaghan (1999). "Conduction Modelling Using Smoothed Particle Hydrodynamics". In: *Journal of Computational Physics* 148.1, pp. 227–264. ISSN: 0021-9991.

Colagrossi, A. (2005). *A meshless Lagrangian method for free-surface and interface flows with fragmentation.* Roma, Italy: La Sapienza.

Colagrossi, A. et al. (2011). "Theoretical analysis and numerical verification of the consistency of viscous smoothed-particle-hydrodynamics formulations in simulating free-surface flows". In: *Physical Review E* 84.2, p. 026705.

Courant, R., K. Friedrichs, and H. Lewy (1928). "Über die partiellen Differenzengleichungen der mathematischen Physik". In: *Mathematische Annalen* 100.1, pp. 32–74.

Crank, J. (1987). *Free and Moving Boundary Problems*. Oxford University Press on Demand. ISBN: 9780198533702.

Cummins, S. J. and M. Rudman (1999). "An SPH Projection Method". In: *Journal of Computational Physics* 152.2, pp. 584–607. ISSN: 0021-9991.

Cundall, P. A. (1971). "A computer model for simulating progressive, large-scale movements in blocky rock systems". In: *Proc. Int. Symp. on Rock Fracture*, pp. 11–8.

Czaputa, K. and G. Brenn (2012). "The convective drying of liquid films on slender wires". In: *International Journal of Heat and Mass Transfer* 55.1–3, pp. 19–31. ISSN: 0017-9310.

Dalrymple, R. A. and B. D. Rogers (2006). "Numerical modeling of water waves with the SPH method". In: *Coastal Hydrodynamics and Morphodynamics Symposium celebrating the academic closing address of Jurjen A. Battjes* 53.2–3, pp. 141–147. ISSN: 0378-3839.

Das, R. and P. W. Cleary (2013). "A mesh-free approach for fracture modelling of gravity dams under earthquake". In: *International Journal of Fracture* 179.1-2, pp. 9–33. ISSN: 0376-9429.

Dehnen, W. and H. Aly (2012). "Improving convergence in smoothed particle hydrodynamics simulations without pairing instability". In: *Monthly Notices of the Royal Astronomical Society* 425.2, pp. 1068–1082. ISSN: 1365-2966.

Domínguez, J. M. et al. (2011). "Neighbour lists in smoothed particle hydrodynamics". In: *International Journal for Numerical Methods in Fluids* 67.12, pp. 2026–2042. ISSN: 1097-0363.

Duda, J. L. et al. (1982). "Prediction of diffusion coefficients for polymer-solvent systems". In: *AIChE Journal* 28.2, pp. 279–285.

Dušička, E., A. N. Nikitin, and I. Lacík (2019). "Propagation rate coefficient for acrylic acid polymerization in bulk and in propionic acid by the PLP-SEC method: experiment and 3D simulation". In: *Polymer Chemistry* 10 (43), pp. 5870–5878.

Eckhard, S. et al. (2014). "Modification of the mechanical granule properties via internal structure". In: *Powder Technology* 258.0, pp. 252–264. ISSN: 0032-5910.

Escobar-Vargas, J., P. Diamessis, and C. Loan (2011). "The numerical solution of the pressure Poisson equation for the incompressible Navier-Stokes equations using a quadrilateral spectral multidomain penalty method". In:

Español, P. and M. Revenga (2003). "Smoothed dissipative particle dynamics". In: *Phys. Rev. E* 67.2, p. 026705.

Ferziger, J. H. and M. Peric (2007). *Numerische Stromungsmechanik.* Berlin: Springer-Verlag Berlin and Heidelberg GmbH & Co. KG.

Fick, A. (1855). "Ueber Diffusion". In: *Annalen der Physik* 170.1, pp. 59–86. ISSN: 1521-3889.

Flory, P. J. (1942). "Thermodynamics of High Polymer Solutions". In: *The Journal of Chemical Physics* 10.1, pp. 51–61.

Franke, K., H.-U. Moritz, and W. Pauer (2017). "Beeinflussung der Eigenschaften von Sprühpolymerisationsprodukten". In: *Chemie Ingenieur Technik* 89.4, pp. 490–495.

Fredenslund, A., J. Gmehling, and P. Rasmussen (1977). *Vapor-liquid equilibria using UNIFAC.* a group-contribution method. Amsterdam: Elsevier. ISBN: 0-444-41621-8.

Fredenslund, A., R. L. Jones, and J. M. Prausnitz (1975). "Group-contribution estimation of activity coefficients in nonideal liquid mixtures". In: *AIChE Journal* 21.6, pp. 1086–1099. ISSN: 1547-5905.

Gingold, R. A. and J. J. Monaghan (1977). "Smoothed particle hydrodynamics: Theory and application to non-spherical stars". In: *Monthly Notices of the Royal Astronomical Society* 181, pp. 375–389. ISSN: 00358711.

Grenier, N. et al. (2009). "An Hamiltonian interface SPH formulation for multi-fluid and free surface flows". In: *Journal of Computational Physics* 228.22, pp. 8380–8393. ISSN: 0021-9991.

Gridnev, A. A. and S. D. Ittel (1996). "Dependence of Free-Radical Propagation Rate Constants on the Degree of Polymerization". In: *Macromolecules* 29.18, pp. 5864–5874. ISSN: 0024-9297.

Gross, J. and G. Sadowski (2001). "Perturbed-Chain SAFT: An Equation of State Based on a Perturbation Theory for Chain Molecules". In: *Ind. Eng. Chem. Res.* 40.4, pp. 1244–1260. ISSN: 0888-5885.

Guermond, J. L., P. Minev, and J. Shen (2006). "An overview of projection methods for incompressible flows". In: *Computer Methods in Applied Mechanics and Engineering* 195.44–47, pp. 6011–6045. ISSN: 0045-7825.

Gunn, R. and G. D. Kinzer (1949). "The Terminal Velocity of Fall for Water Droplets in Stagnant Air". In: *J. Meteor.* 6.4, pp. 243–248. ISSN: 0095-9634.

Handscomb, C. S., M. Kraft, and A. E. Bayly (2009). "A new model for the drying of droplets containing suspended solids after shell formation". In: *Chemical Engineering Science* 64.2, pp. 228–246. ISSN: 0009-2509.

Harten, A. (1983). "High resolution schemes for hyperbolic conservation laws". In: *Journal of Computational Physics* 49.3, pp. 357–393. ISSN: 0021-9991.

Henson, V. E. and U. M. Yang (Apr. 2002). "BoomerAMG: A parallel algebraic multigrid solver and preconditioner". In: *Applied Numerical Mathematics* 41.1, pp. 155–177. ISSN: 0168-9274.

Hernquist, L. and N. Katz (1989). "TREESPH - A unification of SPH with the hierarchical tree method". In: *The Astrophysical Journal Supplement Series* 70, pp. 419–446. ISSN: 0067-0049.

Hirsch, C. (1990). *Numerical computation of internal and external flows*. Wiley series in numerical methods in engineering. Chichester [England] and New York: John Wiley & Sons. ISBN: 9780471923510.

Hirt, C. W. and B. D. Nichols (Jan. 1981). "Volume of fluid (VOF) method for the dynamics of free boundaries". In: *Journal of Computational Physics* 39.1, pp. 201–225. ISSN: 0021-9991.

Hockney, R. W., S. P. Goel, and J. W. Eastwood (1973). "A 10000 particle molecular dynamics model with long range forces". In: *Chemical Physics Letters* 21.3, pp. 589–591. ISSN: 0009-2614.

Hockney, R. W., S. P. Goel, and J. W. Eastwood (1974). "Quiet high-resolution computer models of a plasma". In: *Journal of Computational Physics* 14.2, pp. 148–158. ISSN: 0021-9991.

Hosseini, S. M. and J. J. Feng (2011). "Pressure boundary conditions for computing incompressible flows with SPH". In: *Journal of Computational Physics* 230.19, pp. 7473–7487. ISSN: 0021-9991.

Hu, X. Y. and N. A. Adams (2006). "A multi-phase SPH method for macroscopic and mesoscopic flows". In: *Journal of Computational Physics* 213.2, pp. 844–861. ISSN: 0021-9991.

Hu, X. Y. and N. A. Adams (2007). "An incompressible multi-phase SPH method". In: *Journal of Computational Physics* 227.1, pp. 264–278. ISSN: 0021-9991.

Hu, X. Y. and N. A. Adams (2009). "A constant-density approach for incompressible multi-phase SPH". In: *Journal of Computational Physics* 228.6, pp. 2082–2091. ISSN: 0021-9991.

Huber, M. et al. (Apr. 2016). "On the physically based modeling of surface tension and moving contact lines with dynamic contact angles on the continuum scale". In: *Journal of Computational Physics* 310, pp. 459–477. ISSN: 0021-9991.

Huggins, M. L. (1941). "Solutions of Long Chain Compounds". In: *The Journal of Chemical Physics* 9.5, pp. 440–440.

Hulburt, H. M. and S. Katz (1964). "Some problems in particle technology: A statistical mechanical formulation". In: *Chemical Engineering Science* 19.8, pp. 555–574. ISSN: 0009-2509.

Hutchinson, R. A. (2005). "Free-radical Polymerization: Homogeneous". In: *Handbook of polymer reaction engineering*. Ed. by T. Meyer and J. Keurentjes. Weinheim: Wiley-VCH, pp. 153–212. ISBN: 9783527310142.

Ito, S.-i. and S. Yukawa (2012). "Dynamical scaling of fragment distribution in drying paste". In: *ArXiv e-prints* 1209.6114v1.

Jubelgas, M., V. Springel, and K. Dolag (2004). "Thermal conduction in cosmological SPH simulations". In: *Monthly Notices of the Royal Astronomical Society* 351.2, pp. 423–435. ISSN: 00358711.

Karunasena, H. C. P., W. Senadeera, R. J. Brown, et al. (2014a). "A particle based model to simulate microscale morphological changes of plant tissues during drying". In: *Soft Matter* 10.29, pp. 5249–5268. ISSN: 1744-683X.

Karunasena, H. C. P., W. Senadeera, R. J. Brown, et al. (2014b). "Simulation of plant cell shrinkage during drying – A SPH–DEM approach". In: *Engineering Analysis with Boundary Elements* 44, pp. 1–18. ISSN: 0955-7997.

Karunasena, H. C. P., W. Senadeera, Y. T. Gu, et al. (2014). "A coupled SPH-DEM model for micro-scale structural deformations of plant cells during drying". In: *Applied Mathematical Modelling* 38.15–16, pp. 3781–3801. ISSN: 0307-904X.

Keller, F. (2015). *Simulation of the Morphogenesis of Open–Porous Materials*. Logos Verlag Berlin.

Kieviet, F. and P. J. A. M. Kerkhof (1995). "Measurements of Particle Residence Time Distributions in A Co-Current Spray Dryer". In: *Drying Technology* 13.5-7, pp. 1241–1248. ISSN: 07373937.

Kondo, M. et al. (2007). "Surface Tension Model Using Inter-Particle Force in Particle Method". In: *ASME Conference Proceedings* 2007.42886, pp. 93–98.

Koshizuka, S., A. Nobe, and Y. Oka (1998). "Numerical analysis of breaking waves using the moving particle semi-implicit method". In: *International*

Journal for Numerical Methods in Fluids 26.7, pp. 751–769. ISSN: 1097-0363.

Krüger, M. (2004). *Sprühpolymerisation: Aufbau und Untersuchung von Modellverfahren zur kontinuierlichen Gleichstrom-Sprühpolymerisation.* 1st ed. Berlin: Wissenschaft-und-Technik-Verlag. ISBN: 9783896852045.

Lattanzio, J. C. et al. (1985). "Interstellar Cloud Collisions". In: *Monthly Notices of the Royal Astronomical Society* 215, p. 125. ISSN: 00358711.

Lechner, M. D., K. Gehrke, and E. H. Nordmeier (1996). *Makromolekulare Chemie: Ein Lehrbuch für Chemiker, Physiker, Materialwissenschaftler und Verfahrenstechniker.* 2nd ed. Basel: Birkhäuser. ISBN: 9783764353438.

Lee, E. .-S. et al. (2008). "Comparisons of weakly compressible and truly incompressible algorithms for the SPH mesh free particle method". In: *Journal of Computational Physics* 227.18, pp. 8417–8436. ISSN: 0021-9991.

Lucy, L. B. (1977). "A numerical approach to the testing of the fission hypothesis". In: *Astronomical Journal* 82.12, pp. 1013–1024.

Marchand, A. et al. (2011). "Why is surface tension a force parallel to the interface?" In: *American Journal of Physics* 79.10, pp. 999–1008.

Mazza, M. G. G., L. E. B. Brandão, and G. S. Wildhagen (2003). "Characterization of the Residence Time Distribution in Spray Dryers". In: *Drying Technology* 21.3, pp. 525–538. ISSN: 07373937.

Meister, A. and C. Vömel (2008). *Numerik linearer Gleichungssysteme: Eine Einführung in moderne Verfahren; mit MATLAB-Implementierungen von C. Vömel.* 3., überarb. Aufl. Wiesbaden: Vieweg. ISBN: 9783834804310.

Mezhericher, M., A. Levy, and I. Borde (2009). "Heat and Mass Transfer and Breakage of Particles in Drying Processes". In: *Drying Technology* 27.7, pp. 870–877. ISSN: 07373937.

Mezhericher, M., A. Levy, and I. Borde (2010). "Theoretical Models of Single Droplet Drying Kinetics: A Review". In: *Drying Technology* 28.2, pp. 278–293. ISSN: 07373937.

Monaghan, J. J. (1989). "On the problem of penetration in particle methods". In: *Journal of Computational Physics* 82.1, pp. 1–15. ISSN: 0021-9991.

Monaghan, J. J. (1992). "Smoothed Particle Hydrodynamics". In: *Annual Review of Astronomy and Astrophysics* 30, pp. 543–574. ISSN: 0066-4146.

Monaghan, J. J. (1994). "Simulating Free Surface Flows with SPH". In: *Journal of Computational Physics* 110.2, pp. 399–406. ISSN: 0021-9991.

Monaghan, J. J. (2005). "Smoothed particle hydrodynamics". In: *Reports on Progress in Physics* 68.8, pp. 1703–1759. ISSN: 0034-4885.

Monaghan, J. J. (2012). "Smoothed Particle Hydrodynamics and Its Diverse Applications". In: *Annual Review of Fluid Mechanics* 44.1, pp. 323–346.

Monaghan, J. J. and R. A. Gingold (1983). "Shock simulation by the particle method SPH". In: *Journal of Computational Physics* 52.2, pp. 374–389. ISSN: 0021-9991.

Monaghan, J. J., H. E. Huppert, and M. G. Worster (2005). "Solidification using smoothed particle hydrodynamics". In: *Journal of Computational Physics* 206.2, pp. 684–705. ISSN: 0021-9991.

Morris, J. P. (2000). "Simulating surface tension with smoothed particle hydrodynamics". In: *International Journal for Numerical Methods in Fluids* 33.3, pp. 333–353. ISSN: 1097-0363.

Morris, J. P., P. J. Fox, and Y. Zhu (1997). "Modeling Low Reynolds Number Incompressible Flows Using SPH". In: *Journal of Computational Physics* 136.1, pp. 214–226. ISSN: 0021-9991.

Muginstein, A., M. Fichman, and C. Gutfinger (2001). "Gas absorption in a moving drop containing suspended solids". In: *International Journal of Multiphase Flow* 27.6, pp. 1079–1094. ISSN: 0301-9322.

Mujumdar, A. (2007). *Handbook of industrial drying*. Boca Raton, FL: CRC/Taylor & Francis. ISBN: 978-1-4665-9665-8.

Nair, P. and G. Tomar (2014). "An improved free surface modeling for incompressible SPH". In: *Computers & Fluids* 102.0, pp. 304–314. ISSN: 0045-7930.

Nesic, S. and J. Vodnik (1991). "Kinetics of droplet evaporation". In: *Chemical Engineering Science* 46.2, pp. 527–537. ISSN: 0009-2509.

Nugent, S. and H. A. Posch (2000). "Liquid drops and surface tension with smoothed particle applied mechanics". In: *Physical Review E* 62.4, pp. 4968–4975.

Parti, M. (1994). "MASS TRANSFER BIOT NUMBERS". In: *Periodica Polytechnica Mechanical Engineering* 38.2-3, pp. 109–122.

Pedregosa, F. et al. (2011). "Scikit-learn: Machine Learning in Python". In: *Journal of Machine Learning Research* 12, pp. 2825–2830.

Price, D. J. (2012). "Smoothed particle hydrodynamics and magnetohydrodynamics". In: *Journal of Computational Physics* 231.3, pp. 759–794. ISSN: 0021-9991.

Randles, P. W. and L. D. Libersky (1996). "Smoothed Particle Hydrodynamics: Some recent improvements and applications". In: *Computer Methods in Applied Mechanics and Engineering* 139.1-4, pp. 375–408. ISSN: 0045-7825.

Ranz, W. E. and W. R. Marshall (1952a). "Evaporation from Drops Part I". In: *Chemical Engineering Progress* 48.3, pp. 141–146. ISSN: 0360-7275.

Ranz, W. E. and W. R. Marshall (1952b). "Evaporation from Drops Part II". In: *Chemical Engineering Progress* 48.4, pp. 173–180. ISSN: 0360-7275.

Rasmussen, C. E. and C. K. I. Williams (2006). *Gaussian Processes for Machine Learning*. MIT Press. ISBN: 026218253X.

Ray, W. H. (1972). "On the Mathematical Modeling of Polymerization Reactors". In: *Polymer Reviews* 8.1, pp. 1–56.

Rosswog, S. (2009). "Astrophysical smooth particle hydrodynamics". In: *New Astronomy Reviews* 53.4-6, pp. 78–104. ISSN: 1387-6473.

Ryan, E. M., A. M. Tartakovsky, and C. Amon (2010). "A novel method for modeling Neumann and Robin boundary conditions in smoothed particle hydrodynamics". In: *Computer Physics Communications* 181.12, pp. 2008–2023. ISSN: 0010-4655.

Schofield, P. (1973). "Computer simulation studies of the liquid state". In: *Computer Physics Communications* 5.1, pp. 17–23. ISSN: 0010-4655.

Sethian, J. (1999). *Level set methods and fast marching methods : evolving interfaces in computational geometry, fluid mechanics, computer vision, and materials science*. Cambridge, U.K. New York: Cambridge University Press. ISBN: 0-5216-455-73.

Seydel, P. (2005). "Modellierung der Feststoffbildung in Einzeltropfen bei der Sprühtrocknung". PhD thesis. Bochum: Ruhr-Universität.

Shao, S. and E. Y. M. Lo (2003). "Incompressible SPH method for simulating Newtonian and non-Newtonian flows with a free surface". In: *Advances in Water Resources* 26.7, pp. 787–800. ISSN: 0309-1708.

Sloth, J. et al. (2006). "Model based analysis of the drying of a single solution droplet in an ultrasonic levitator". In: *Chemical Engineering Science* 61.8, pp. 2701–2709. ISSN: 0009-2509.

Stellingwerf, R. F. and C. A. Wingate (1994). "Impact Modelling with Smoothed Particle Hydrodynamics". In: *Memorie della Societa Astronomica Italiana Supplementi* 65, p. 1117.

Stokes, G. G. (1851). "On the Effect of the Internal Friction of Fluids on the Motion of Pendulums". In: *Transactions of the Cambridge Philosophical Society* 9, pp. 8–106.

Szewc, K., J. Pozorski, and J.-P. Minier (2012). "Analysis of the incompressibility constraint in the smoothed particle hydrodynamics method". In: *International Journal for Numerical Methods in Engineering* 92.4, pp. 343–369. ISSN: 1097-0207.

Takeda, H., S. M. Miyama, and M. Sekiya (1994). "Numerical Simulation of Viscous Flow by Smoothed Particle Hydrodynamics". In: *Progress of Theoretical Physics* 92.5, pp. 939–960.

Tartakovsky, A. M. and P. Meakin (2005). "Modeling of surface tension and contact angles with smoothed particle hydrodynamics". In: *Physical Review E* 72.2, p. 026301.

Taylor, R. and R. Krishna (1993). *Multicomponent mass transfer*. New York: Wiley. ISBN: 9780471574170.

Vacondio, R. et al. (Sept. 2020). "Grand challenges for Smoothed Particle Hydrodynamics numerical schemes". In: *Computational Particle Mechanics*. ISSN: 2196-4386.

van Leer, B. (1973). "Towards the ultimate conservative difference scheme I. The quest of monotonicity". In: *Proceedings of the Third International Conference on Numerical Methods in Fluid Mechanics*. Ed. by H. Cabannes and R. Temam. Vol. 18. Lecture Notes in Physics. Springer Berlin Heidelberg, pp. 163–168. ISBN: 978-3-540-06170-0.

Verlet, L. (1967). "Computer "Experiments" on Classical Fluids. I. Thermodynamical Properties of Lennard-Jones Molecules". In: *Physical Review* 159.1, pp. 98–103.

Virtanen, P. et al. (2020). "SciPy 1.0: Fundamental Algorithms for Scientific Computing in Python". In: *Nature Methods* 17, pp. 261–272.

Vrentas, J. S., J. L. Duda, and H.-C. Ling (1985). "Free-volume theories for self-diffusion in polymer–solvent systems. I. Conceptual differences in theories". In: *Journal of Polymer Science: Polymer Physics Edition* 23.2, pp. 275–288. ISSN: 1542-9385.

Vrentas, J. S., J. L. Duda, H.-C. Ling, and A.-C. Hou (1985). "Free-volume theories for self-diffusion in polymer–solvent systems. II. Predictive capabilities". In: *Journal of Polymer Science: Polymer Physics Edition* 23.2, pp. 289–304. ISSN: 1542-9385.

Walag, K. (2011). *Sprühpolymerisation: Modellierung der Mikropolymerisation im Turmreaktor.* 1st ed. Berlin: Wissenschaft-und-Technik-Verlag. ISBN: 3896852264.

Watkins, S. J. et al. (1996). "A new prescription for viscosity in Smoothed Particle Hydrodynamics". In: *Astron. Astrophys. Suppl. Ser.* 119.1, pp. 177–187.

Wessing, S. (2015). *Two-stage methods for multimodal optimization.* en. Technische Universität Dortmund.

Wittenberg, N. F. G. (2013). *Kinetics and Modeling of the Radical Polymerization of Acrylic Acid and of Methacrylic Acid in Aqueous Solution.* Göttingen: Niedersächsische Staats- und Universitätsbibliothek Göttingen.

Woog, T. (2011). "Partikelbasierte Simulationen zur Fluiddynamik unter Betrachtung von freien Oberflächen und Oberflächenspannung mit SPH". MA thesis. Universität Stuttgart, Institut für Chemische Verfahrenstechnik.

Wulkow, M. (2008). "Computer Aided Modeling of Polymer Reaction Engineering—The Status of Predici, I-Simulation". In: *Macromolecular Reaction Engineering* 2.6, pp. 461–494.

Yeo, I.-K. and R. A. Johnson (Jan. 2000). "A New Family of Power Transformations to Improve Normality or Symmetry". In: *Biometrika* 87.4, pp. 954–959. ISSN: 00063444.

Zhu, Y. and P. J. Fox (2001). "Smoothed Particle Hydrodynamics Model for Diffusion through Porous Media". In: *Transport in Porous Media* 43.3, pp. 441–471.